Chemistry Units 3 & 4

First published in 2024
Insight Publications Pty Ltd
3/350 Charman Road
Cheltenham Victoria 3192
Australia

Tel: +61 3 8571 4950
Email: books@insightpublications.com.au

www.insightpublications.com.au

Reproduction and communication for educational purposes:

Reproduction and communication for other purposes:

Insight VCE Revision Questions: Chemistry Units 3&4

ISBN: 9781923154926

VCE is a registered trademark. The VCAA does not endorse or make any warranties regarding this study resource. Current VCE Study Designs, exam specifications and past VCE exams can be accessed directly at www.vcaa.vic.edu.au .

Written by Pat O'Shea
Reviewed by Mark Mitchell
Edited by Anna Alberti
Proofread by Geoffrey Marnell
Cover by Melisa Paredes
Internal design and layout by Bec Yule @ Red Chilli Design and Melisa Paredes
Printed by Markono Print Media Pte Ltd

Insight Publications acknowledges the Traditional Custodians of the Country on which we meet and work, the Boonwurrung People of the Kulin Nation. We pay our respects to their Elders past and present, and extend that respect to all Aboriginal and Torres Strait Islander peoples.

Contents

Introduction   iv

Questions   1

    Unit 3, Area of Study 1: What are the current and future options for supplying energy?   1

    Unit 3, Area of Study 2: How can the rate and yield of chemical reactions be optimised?   17

    Unit 4, Area of Study 1: How are organic compounds categorised and synthesised?   47

    Unit 4, Area of Study 2: How are organic compounds analysed and used?   58

    Unit 4, Area of Study 3: How is scientific inquiry used to investigate the sustainable production of energy and/or materials?   84

Worked solutions   90

    Unit 3, Area of Study 1: What are the current and future options for supplying energy?   90

    Unit 3, Area of Study 2: How can the rate and yield of chemical reactions be optimised?   109

    Unit 4, Area of Study 1: How are organic compounds categorised and synthesised?   142

    Unit 4, Area of Study 2: How are organic compounds analysed and used?   155

    Unit 4, Area of Study 3: How is scientific inquiry used to investigate the sustainable production of energy and/or materials?   182

Introduction

Insight's *VCE Revisions Questions: Chemistry Units 3 & 4* contains questions, answers, explanatory notes and tips. A good habit to implement is to test yourself by working through this resource. The process of actively recalling information assists with deeper learning, and it helps that you can check whether your answer is correct.

By using this resource as part of your study regime throughout the year you will be prepared for questions you may encounter in your VCE exam.

We wish you well with your studies.

The Insight Team

● Questions

Unit 3 | Area of Study 1   What are the current and future options for supplying energy?

Multiple choice

Question 1

Which fuel is the most sustainable?

A.   natural gas obtained from fracking for coal

B.   ethanol obtained from ethane in natural gas

C.   bioethanol produced from sugar cane

D.   biodiesel produced from whale blubber

Question 2

Which one of the following statements about the use of canola crops to make biodiesel is the most correct?

A.   The large-scale growing of canola crops in Australia presents environmental challenges.

B.   The whole canola plant is used in the production of biodiesel.

C.   Canola oil undergoes a condensation reaction with methanol to form an ester.

D.   Each triglyceride molecule produces one biodiesel molecule.

Question 3

In emergency situations, diesel generators are often transported to areas that have lost electrical power. A diesel generator is a compact unit that converts the potential energy of diesel into electrical energy.

The main reason diesel generators are used for this purpose is that they

A.   produce low levels of emissions.

B.   use fuel that is renewable.

C.   are the most efficient way of producing electrical energy.

D.   produce electrical energy very quickly after start-up.

Question 4

A typical component of diesel is the alkane dodecane. Each molecule of dodecane contains 38 atoms.

The molecular formula of dodecane is

A. $C_{12}H_{26}$

B. $C_{13}H_{26}$

C. $C_{18}H_{38}$

D. $C_{38}H_{78}$

Question 5

An energy profile diagram for a reaction is shown below.

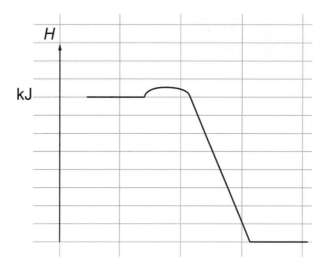

This energy profile diagram represents

A.   very stable reactants that will require significant amounts of energy to cause a reaction.

B.   very stable reactants that will release significant amounts of energy upon reaction.

C.   very unstable reactants that will release small amounts of energy upon reaction.

D.   very unstable reactants that will release significant amounts of energy upon reaction.

Question 6

Consider the energy profile diagram below. The enthalpies of the reactants and the products are shown on the vertical axis.

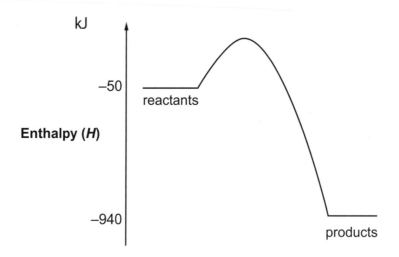

Which fuel has this energy profile diagram when one mole undergoes complete combustion?

A.   hydrogen

B.   methane

C.   methanol

D.   ethanol

Question 7

The energy profile diagrams below compare the combustion of methane and the combustion of Substance B.

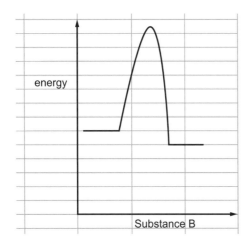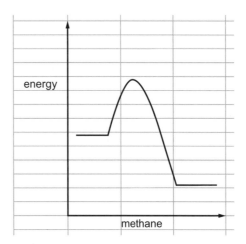

It is likely that Substance B

A.   would be a viable fuel because its combustion is exothermic.

B.   would not be a viable fuel because its low activation energy will make it unstable and dangerous.

C.   would not be a viable fuel because the heat of combustion is too low.

D.   would not be a viable fuel because its combustion reaction is endothermic.

Question 8

Which one of the following releases the greatest amount of energy during complete combustion?

A.   50 g of methane gas

B.   99.2 L of methane gas stored at standard laboratory conditions (SLC)

C.   3.4 mol of methane gas

D.   50 g of ethane gas

Question 9

The properties of a particular fuel are listed in the following table.

boiling point	97.2 °C
heat of combustion	33.7 kJ g^{-1}
solubility in water	high

The fuel is most likely

A.   pentane.

B.   propan-1-ol.

C.   cyclohexane.

D.   stearic acid.

Question 10

What mass of triglyceride is needed to produce the same amount of energy as 1.00 g of the carbohydrate starch?

A. 0.200 g

B. 0.432 g

C. 1.00 g

D. 2.31 g

Question 11

The waste from a sugar refinery is collected and allowed to settle in large vats, as shown in the diagram below.

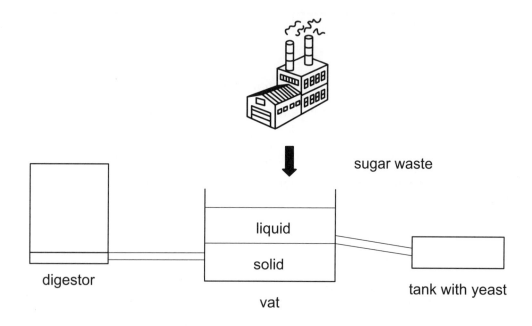

The liquid is tapped from the vats and added to a tank that contains yeast. The solid slurry from the vats is added to an anaerobic digestor that contains bacteria.

As a result of this processing, the refinery is able to produce

A. biogas from both tanks.

B. bioethanol from both tanks.

C. biogas from one tank and bioethanol from the other.

D. both bioethanol and biogas from each tank.

> *Use the following information to answer Questions 12–15.*
>
> Butane is a significant component of liquified petroleum gas (LPG). The equation for the complete combustion of butane is $2C_4H_{10}(g) + 13O_2(g) \rightarrow 8CO_2(g) + 10H_2O(g)$.

Question 12

The energy, in kJ, released from the combustion of 1.00 g of butane will be

A. 49.7

B. 58.0

C. 2880

D. 49700

Question 13

The mass, in g, of O_2 gas required for the complete combustion of 1.16 g of butane will be

A. 2.08

B. 4.16

C. 8.32

D. 16.0

Question 14

The volume, in L, of CO_2 gas produced from the complete combustion of 1.16 g of butane at SLC will be

A. 0.0037

B. 0.149

C. 0.50

D. 1.98

Question 15

What volume of ethane at standard laboratory conditions (SLC) is required to produce the same amount of energy as 150 g of butane?

A. 24.8 L

B. 119 L

C. 1560 L

D. 7460 L

Question 16

The calibration factor of a calorimeter is determined to be 586 J $°C^{-1}$. A 1.80 g sample of cashew is burnt under this calorimeter. The heat of combustion of cashew is quoted as 5.4 kJ g^{-1}.

The expected rise in temperature, in °C, in the calorimeter will be

A. 8.3

B. 16.6

C. 22.6

D. 33.2

Question 17

The food label below is taken from a packet of muesli bars.

Nutrition information			
Servings per package: 6		Serving size: 35 g	
	Average quantity per serve (35 g)	% DI** per serve	Average quantity per 100 g
Energy	555 kJ	6%	1590 kJ
	133 cal		379 cal
Protein	3.8 g	8%	10.8 g
Fat, total	4.9 g	7%	14.0 g
-saturated	0.6 g	3%	1.8 g
Carbohydrate	12.8 g	4%	36.6 g

1.0 g of muesli bar is burnt under a can containing 500 g of water.

Assuming that the energy transfer from the muesli bar to the water is 100% efficient, the temperature change of the water will be

A.   2.7 °C

B.   7.6 °C

C.   15 °C

D.   38 °C

Use the following information to answer Questions 18 and 19.

A student places a coil in a steel can and uses the coil to electrically calibrate the steel can. He then burns a biscuit sample under the steel can. The data from the experiment is shown below.

Mass of biscuit: 1.12 g

Water temperature before biscuit is burnt: 18.8 °C

Water temperature after biscuit is burnt: 29.6 °C

Calibration:

- Voltage 5.40 V, current 3.40 A, time 5.00 min
- Temperature of water before: 21.2 °C
- Temperature of water after: 27.6 °C

Question 18

In J °C^{-1}, the calibration factor for the calorimeter will be

A.   14.3

B.   86.1

C.   861

D.   1720

Question 19

The heat of combustion of the biscuit will be

A. 212 J g^{-1}

B. 8.30 kJ g^{-1}

C. 9.30 kJ g^{-1}

D. 9.30 kJ mol^{-1}

Question 20

Which one of the following options lists sulfur compounds in order of lowest oxidation state of sulfur to highest?

A. H_2S, SO_2, SO_3, S_8

B. H_2S, S_8, SO_2, SO_3

C. SO_3, SO_2, S_8, MgS

D. SO_2, S_8, MgS, SO_3

Question 21

The equation for the redox reaction between manganese dioxide and hydrochloric acid is
$MnO_2(s) + 4HCl(aq) \rightarrow MnCl_2(aq) + 2H_2O(l) + Cl_2(g)$.

The reduction half-equation occurring is

A. $2Cl^-(aq) \rightarrow Cl_2(g) + 2e^-$

B. $4H^+(aq) + 4e^- \rightarrow 2H_2(g)$

C. $MnO_2(s) + 4HCl(aq) \rightarrow MnCl_2(aq) + 2H_2O(l) + 2Cl_2(g)$

D. $MnO_2(s) + 4H^+(aq) + 2e^- \rightarrow Mn^{2+}(aq) + 2H_2O(l)$

Question 22

Ethanol can be oxidised in acidic conditions to ethanoic acid.

The half-equation for this reaction is

A. $C_2H_5O(aq) + H_2O(l) \rightarrow C_2H_4O_2(aq) + 3H^+(aq) + 4e^-$

B. $C_2H_6O(aq) + H_2O(l) + 4e^- \rightarrow C_2H_4O_2(aq) + 4H^+(aq)$

C. $C_2H_6O(aq) + H_2O(l) \rightarrow C_2H_4O_2(aq) + 4H^+(aq) + 4e^-$

D. $C_2H_6O(aq) + OH^-(aq) \rightarrow C_2H_4O_2(aq) + H_2(g) + 2e^-$

Question 23

A galvanic cell is constructed from standard iodine and aluminium half-cells.

In this cell

A.   aluminium will be the reducing agent and electrons will flow to the iodine terminal.

B.   aluminium ions are the oxidising agent and aluminium is the negative terminal.

C.   iodine is oxidised and aluminium ions are reduced.

D.   iodide ions will be oxidised and aluminium will be the positive terminal.

Question 24

A description of a new cell is given below.

'This cell features a continuous supply of fuel and air. It is very expensive but offers a high efficiency and potentially sustainable fuel.'

The cell reactants are most likely to be

A. PbO_2 and Pb.

B. aluminium and oxygen.

C. methane and oxygen.

D. butane and oxygen.

Question 25

The correct half equation for the reaction of IO_4^- to IO_3^- is

A. $IO_4^-(aq) + 2H^+(aq) + 2e^- \rightarrow IO_3^-(aq) + H_2O(l)$

B. $IO_4^-(aq) + 2H^+(aq) + e^- \rightarrow IO_3^-(aq) + H_2O(l)$

C. $IO_4^- + 2H_2O(aq) + 2e^- \rightarrow IO_3^-(aq) + 4H^+(l)$

D. $IO_4^- + 2H^+(aq) \rightarrow 2e^- + IO_3^-(aq) + H_2O(l)$

Question 26

Which of the following is a correct application of the electrochemical series provided in the Data Book?

A. The difference in potential of the reactants indicates the likely reaction rate.

B. Some species appear in more than one half-equation on the electrochemical series.

C. The cell voltage will always be the difference in potential of the reactants.

D. The electrochemical series includes every potential redox reactant.

Question 27

Methane can be used as a reactant in a fuel cell.

If the cell is operating in acidic conditions, the reaction occurring at the anode will be

A. $O_2(g) + 4H^+(aq) + 4e^- \rightarrow 2H_2O(g)$

B. $CH_4(g) + 2O^{2-}(aq) \rightarrow CO_2(g) + 2H_2O(g) + 4e^-$

C. $CH_4(g) + 2H_2O(l) \rightarrow CO_2(g) + 8H^+(aq) + 8e^-$

D. $CH_4(g) + 2O_2(g) \rightarrow CO_2(g) + 2H_2O(g)$

Question 28

The half-equation for the reaction at the cathode of an ethane–oxygen alkaline fuel cell is

A. $O_2(g) + 2H_2O(l) + 4e^- \rightarrow 4OH^-(aq)$

B. $O_2(g) + 4H^+(aq) + 4e^- \rightarrow 2H_2O(l)$

C. $C_2H_6(g) + 14OH^-(aq) \rightarrow 2CO_2(g) + 10H_2O(l) + 14e^-$

D. $C_2H_6(g) + 4H_2O(l) \rightarrow 2CO_2(g) + 14H^+(aq) + 14e^-$

Short answer

Question 1 (7 marks)

a. Wastewater treatment plants produce a thick sludge that is regularly added to anaerobic digestors to produce a renewable biofuel.

    i. Define the term 'renewable' in reference to fuels.   1 mark

    ii. What does 'anaerobic' mean?   1 mark

    iii. Identify the main component of this fuel.   1 mark

    iv. As this fuel is renewable, will it produce harmful emissions?   1 mark

b. i. Which contains the greater chemical potential energy: 1 mole of glucose and 6 moles of oxygen, or 6 moles of CO_2 and 6 moles of H_2O?   1 mark

    ii. In living organisms, where is CO_2 converted to glucose?   1 mark

    iii. In living organisms, where is glucose converted to CO_2?   1 mark

Question 2 (8 marks)

a.   Ethanol can be produced from carbohydrates, such as glucose.

    i.   Write a balanced equation for the production of ethanol from glucose.   1 mark

    ii.   Is the ethanol produced this way considered to be bioethanol?
          Explain your answer.   1 mark

b.   i.   Write a balanced equation for the complete combustion of ethanol in air.   1 mark

    ii.   Calculate the amount of energy released by the complete combustion of
          10.0 kg of ethanol.   2 marks

c.   i.   Write a balanced equation for the incomplete combustion of ethanol to
          form carbon monoxide and water.   1 mark

    ii.   Calculate the volume of CO that will be formed at standard laboratory
          conditions (SLC) from the incomplete combustion of 10.0 kg of ethanol.   2 marks

Question 3 (7 marks)

a. The activation energy required for the conversion of 1 mole of CO_2 and water to methane and oxygen is 1440 kJ.

    i. Write a balanced equation for the complete combustion of 1 mole of methane gas.   1 mark

    ii. Use the axis provided, and the initial enthalpy value, to complete the energy profile diagram for this reaction.   2 marks

b. Natural gas is used in Australia as a fuel in many buses and trucks. The density of liquid natural gas is 0.66 g mL^{-1}.

    i. Assuming natural gas is pure methane, determine the energy that could be produced from a 50.0 L sample of liquid methane.   1 mark

    ii. Calculate the volume of the CO_2 produced at SLC from the combustion of 50.0 L of methane.   3 marks

Question 4 (6 marks)

The energy content of a biscuit can be determined by burning a sample under a beaker of water. The apparatus used is shown in the diagram below.

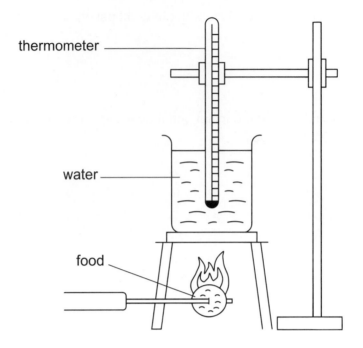

The results obtained for a particular experiment are shown below.

- Mass of biscuit before heating: 3.782 g
- Mass of biscuit after combustion: 1.122 g
- Volume of water: 900.0 mL
- Initial temperature of water: 18.8 °C
- Final temperature of water: 24.4 °C

a. i.   Calculate the energy released by the biscuit.   2 marks

ii.   Calculate the heat of combustion of the biscuit, in kJ g^{-1}.   1 mark

b. The heat of combustion value obtained is significantly lower than the value on the packet label.

Suggest three modifications to the experiment design that should lead to a more accurate result for the heat of combustion.   3 marks

_____

_____

_____

_____

_____

Question 5 (8 marks)

An Australian company is researching the production of a biodiesel molecule, which is shown below.

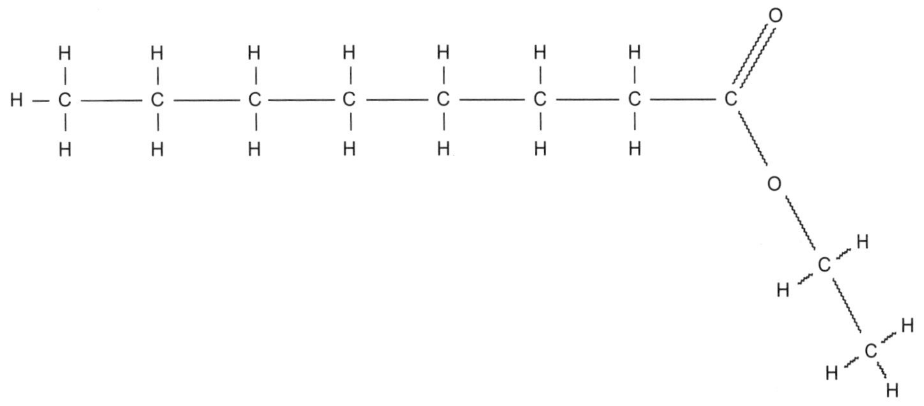

Biodiesel A

Interest in Biodiesel A stems from the fact that the fatty acid required is caprylic acid, a component of cow's milk that is present in waste from cheese-making processes.

a. The alcohol molecule used to make Biodiesel A is not the usual choice of alcohol when making biodiesel.

Explain how the choice of the alcohol used could reduce the environmental impact of making this form of biodiesel.   2 marks

_____

_____

_____

_____

b. Write a balanced equation for the complete combustion of this biodiesel molecule.   2 marks

_____

c. Determine the mass of CO_2 that would be produced from the complete combustion of 1.00 kg of this biodiesel.   4 marks

_____

_____

_____

_____

Question 6 (5 marks)

A biogas generator is built at a metropolitan sewage facility. The gas generated is purified and passed into a methane fuel cell to produce electrical energy.

a. i. What is biogas?   1 mark

_____

ii. Explain why biogas is considered a renewable fuel.   1 mark

_____

b. Assuming that the fuel cell is operating in alkaline conditions, write balanced equations or half-equations for the

i. reaction at the anode.   1 mark

_____

ii. reaction at the cathode.   1 mark

_____

iii. overall equation.   1 mark

_____

Question 7 (5 marks)

Silver and zinc can be used in button batteries. The overall equation is

$$Zn(s) + Ag_2O(s) + H_2O(l) \rightarrow Zn(OH)_2(s) + 2Ag(s)$$

a. i.   Give the oxidation state change of the silver during discharge.   1 mark

ii.   State whether the silver oxide is the anode or the cathode.   1 mark

iii.   State which electrode is the positive electrode.   1 mark

b.   Describe the changes you would observe over time at the

i.   anode.   1 mark

ii.   cathode.   1 mark

Question 8 (7 marks)

Lithium is the key element in many new cells. One example is the lithium–iron cell. It is of interest because the voltage produced is 1.5 V, making it compatible with the conventional cells used in many common appliances. The lithium-iron cell can produce 2.5 times the capacity of an alkaline cell.

The overall equation for this cell is

$$FeS_2 + 4Li \rightarrow Fe + 2Li_2S$$

Phases are not shown in this equation because an organic solvent is used.

a.   Write half-equations in the spaces provided and indicate the polarity of each electrode.

(Use the likely oxidation state of lithium to deduce the oxidation states of the other elements.)   4 marks

	Half-equation	Polarity
anode:	_____	_____
cathode:	_____	_____

b.   Explain why an organic solvent is used in these cells.   1 mark

_____

c.   Explain why a voltage of 1.5 V might be considered advantageous.   1 mark

_____

d.   This cell will eventually go flat. In terms of the chemicals involved, explain why this happens.   1 mark

_____

Unit 3 | Area of Study 2 — How can the rate and yield of chemical reactions be optimised?

Multiple choice

Question 1

A wide variety of lithium cells exist, as manufacturers attempt to find substances that are compatible with lithium metal. One commercial version uses the reaction between lithium metal and sulfur to produce electrical energy. The reaction can be represented by the following equation.

$$16Li + S_8 \rightarrow 8Li_2S$$

The half-equation for the reaction of sulfur in this cell is

A. $S \rightarrow S^{2-} + 2e^-$

B. $S_8 + 2e^- \rightarrow 8S^{2-}$

C. $S_8 \rightarrow 8S^{2-} + 8e^-$

D. $S_8 + 16e^- \rightarrow 8S^{2-}$

Use the following information to answer Questions 2–4.

Lithium and manganese dioxide are used in a commercial secondary cell. The cell is relatively inexpensive and produces a voltage of 3.0 V.

The cathode reaction in this cell is

$$MnO_2(s) + Li^+ + e^- \rightarrow LiMnO_2(s)$$

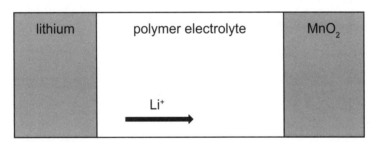

Question 2

The overall equation in this cell during discharge will be

A. $MnO_2(s) + Li^+ + Li \rightarrow Li_2MnO_2(s)$

B. $Mn(s) + Li^+ + O_2 \rightarrow LiMnO_2(s)$

C. $MnO_2(s) + Li^+ \rightarrow Li(s) + MnO_2(s)$

D. $Li(s) + MnO_2(s) \rightarrow LiMnO_2(s)$

Question 3

During discharge, the oxidation state change at the cathode is

A. $Li(s)$ to Li^+

B. Mn^{4+} to Mn^{2+}

C. Mn^{4+} to Mn^{3+}

D. Mn^{4+} to $Mn(s)$

Question 4

When this cell is being recharged, the half-equation occurring at the anode will be

A. $MnO_2(s) + Li^+ + e- \rightarrow LiMnO_2(s)$

B. $LiMnO_2(s) \rightarrow MnO_2(s) + Li^+ + e^-$

C. $Li^+ + e^- \rightarrow Li(s)$

D. $Li(s) \rightarrow Li^+ + e^-$

> *Use the following information to answer Questions 5 and 6.*
>
> The overall equation for a lead–acid accumulator is
>
> $$Pb(s) + PbO_2(s) + 2H_2SO_4(aq) \rightarrow 2PbSO_4(s) + 2H_2O(l)$$

Question 5

As this cell discharges, the

A.   pH of the battery will rise to approximately 7.

B.   mass of the cell will decrease.

C.   products will move into the solution in the cell.

D.   lead atoms will all change from Pb^{4+} to Pb^{2+}.

Question 6

The main reason this cell can be recharged is that the

A.   reaction is occurring in an aqueous environment.

B.   voltage produced by the cell is not relatively high.

C.   products of the reaction remain in contact with the electrodes.

D.   same element is reacting at both electrodes.

Use the following information to answer Questions 7 and 8.

The diagram shows a cross-section of a zinc–bromine flow battery.
This cell can be charged to produce zinc and bromine. When power is required, the zinc and bromine can be reacted with each other.

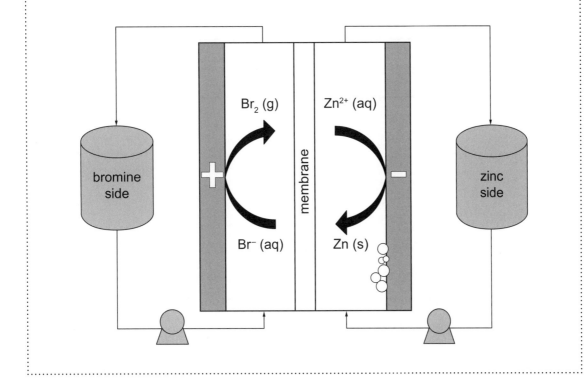

Question 7

In this cell, whether charging or discharging, the

A. zinc will always be the negative electrode.

B. bromine will always be the anode.

C. electrons will always flow from the zinc electrode to the bromine electrode.

D. zinc will always be the anode.

Question 8

The overall equation in this cell when it is charging will be

A. $Zn(s) + Br_2(l) \rightarrow Zn^{2+}(aq) + 2Br^-(aq)$

B. $Zn^{2+}(aq) + Br_2(l) \rightarrow Zn(s) + 2Br^-(aq)$

C. $Zn^{2+}(aq) + 2Br^-(aq) \rightarrow Zn(s) + Br_2(l)$

D. $Zn(s) + 2Br^-(aq) \rightarrow Zn^{2+}(aq) + Br_2(l)$

Use the following information to answer Questions 9 and 10.

The diagram below shows a representation of a lithium–manganese dioxide cell.

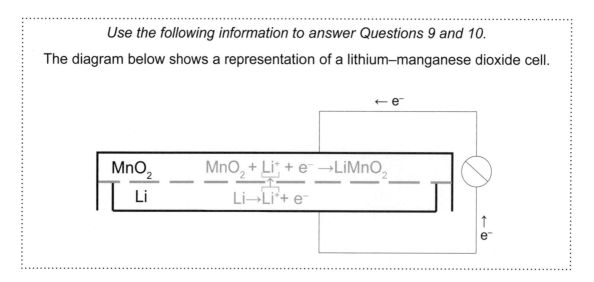

Question 9

In this cell

A. electrons will flow from the cathode to the anode.

B. the lithium metal is the cathode and has a positive polarity.

C. lithium metal is oxidised and manganese ions are reduced.

D. lithium ions are reduced and manganese ions are oxidised.

Question 10

The overall reaction (simplified) for this cell will be

A. $Li(s) + MnO_2(s) \rightarrow LiMnO_2(s) + Li^+$

B. $Li(s) + MnO_2(s) \rightarrow LiMnO_2(s)$

C. $Li^+ + MnO_2(s) + Li(s) \rightarrow Li^+ + LiMnO_2(s)$

D. $Li^+ + MnO_2(s) \rightarrow LiMnO_2(s)$

Question 11

The equation for the decomposition of hydrogen peroxide is

$$2H_2O_2(l) \rightarrow 2H_2O(l) + O_2(g)$$

This is a relatively slow reaction that can be catalysed by the addition of manganese dioxide or by the addition of pieces of liver from animals. There is an enzyme in liver that catalyses the reaction.

Two experiments with hydrogen peroxide and MnO_2 are conducted.

Experiment 1	Experiment 2
100 mL of H_2O_2 with 1 spatula of MnO_2	100 mL of H_2O_2 with 2 spatulas of MnO_2

Both reactions are monitored until no further reaction occurs.

Select the correct statement comparing the two experiments.

A. Experiment 2 will produce twice the volume of oxygen gas.

B. The volume of oxygen gas will depend upon which chemical is in excess.

C. Both reactions produce the same volume of oxygen gas but at different rates.

D. The rate of reaction will be the same in both flasks.

Question 12

A student wishes to investigate the rate of reaction between calcium carbonate and hydrochloric acid. She prepares five HCl solutions of concentrations ranging between 0.1 M to 0.5 M. She adds 50 mL of each solution to a different beaker and places the five beakers on a hot plate set to 50 °C. Once the contents of the beaker have reached 50 °C, she adds 1.0 g of calcium carbonate to each beaker and records the time required for the reaction to cease.

Select the option that correctly identifies the variables in this experiment.

	Control variable	Independent variable	Dependent variable
A.	temperature	time for reaction	HCl concentration
B.	temperature	HCl concentration	time for reaction to cease
C.	HCl concentration	mass of $CaCO_3$	time for reaction to cease
D.	time for reaction	HCl concentration	mass of $CaCO_3$

Use the following information to answer Questions 13 and 14.

A student performs a series of experiments during which they add hydrochloric acid to calcium carbonate, $CaCO_3$ (Mr = 100.1 g mol^{-1}). With each experiment, the student monitors the volume of carbon dioxide gas released. The results of the experiment are shown on the graphs below.

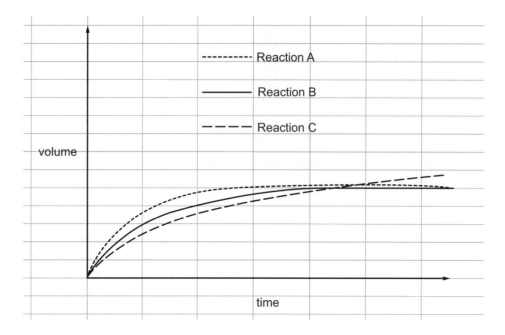

Reaction B involves 1.0 g of small marble chips being added to 100 mL of 1.0 M HCl.

Question 13

Which one of the following statements is consistent with the graph for Reaction A?

A. The only change made to the procedure for Reaction B was to lower the temperature of the HCl.

B. The mass of calcium carbonate used in Reaction A was greater than that used in Reaction B.

C. The concentration of the HCl in Reaction A was higher than that in Reaction B.

D. The shape of the beaker used in Reaction A was different from that used in Reaction B.

Question 14

Which one of the following statements is consistent with the graph for Reaction C?

A. A catalyst was added to Reaction C along with the calcium carbonate.

B. A single marble chip of mass 1.2 g was added to the HCl in Reaction C.

C. The 1.0 g of calcium carbonate was added as a single marble chip in Reaction C.

D. The concentration of the HCl was slightly lower in Reaction C than in Reaction B.

Question 15

An expression for an equilibrium constant is shown below.

$$K = \frac{[H_2]^{\frac{1}{2}}[I_2]^{\frac{1}{2}}}{[HI]}$$

This expression is for the equation

A. $2HI(g) \rightleftharpoons H_2(g) + I_2(g)$

B. $\frac{1}{2}H_2(g) + \frac{1}{2}I_2(g) \rightleftharpoons HI(g)$

C. $H_2(g) + I_2(g) \rightleftharpoons 2HI(g)$

D. $HI(g) \rightleftharpoons \frac{1}{2}H_2(g) + \frac{1}{2}I_2(g)$

Use the following information to answer Questions 16 and 17.

The reaction between methane and steam can be used to produce hydrogen gas. The equation for this reaction is shown below.

$$CH_4(g) + 2H_2O(g) \rightleftharpoons CO_2(g) + 4H_2(g), \Delta H < 0$$

Question 16

The units for the equilibrium constant, K, for this reaction are

A.   M

B. M^{-1}

C. M^2

D. M^{-2}

Question 17

Which one of the following changes would result in an increased yield of hydrogen gas?

A.   an increase in temperature

B.   a decrease in pressure

C.   a decrease in volume

D.   the addition of extra CO_2

Question 18

The reaction between nitrogen and hydrogen gases to produce ammonia is

$$N_2(g) + 3H_2(g) \rightleftharpoons 2NH_3(g), \Delta H = -91 \text{ kJ}$$

Which one of the following changes to an equilibrium mixture of these gases will lead to a greater yield of ammonia?

A.   an increase in temperature

B.   an increase in volume

C.   the addition of a catalyst

D.   an increase in pressure

Question 19

Ammonia is manufactured through the reaction between nitrogen and hydrogen gases. The equation for this exothermic, reversible reaction is shown below.

$$N_2(g) + 3H_2(g) \rightleftharpoons 2NH_3(g)$$

The temperature of an equilibrium mixture of these gases is increased. Which option is consistent with an increase in temperature of the system?

	Amount N_2	Amount H_2	Amount NH_3
A.	increase 0.4 mol	increase 1.2 mol	decrease 0.8 mol
B.	increase 0.4 mol	increase 0.4 mol	decrease 0.4 mol
C.	increase 0.4 mol	increase 1.2 mol	increase 0.8 mol
D.	decrease 0.4 mol	decrease 1.2 mol	increase 0.8 mol

> *Use the following information to answer Questions 20 and 21.*
>
> One of the steps in the manufacturing of nitric acid is the reaction between ammonia and oxygen. The reaction can be represented by the following equation.
>
> $$4NH_3(g) + 5O_2(g) \rightleftharpoons 4NO(g) + 6H_2O(g), \Delta H = -ve$$

Question 20

The units for the equilibrium constant, K, for this reaction are

A. M^{-2}

B. M^{-1}

C. M

D. M^2

Question 21

The volume of an equilibrium mixture of the above gases is halved at time t_1, and the mixture regains equilibrium at time t_2.

Which one of the following statements best describes the system?

A. The concentration of NO at t_2 will be higher than it was at t_1.

B. The amount of NO at t_2 will be higher than it was at t_1.

C. The amount of NO will not change between t_1 and t_2 because temperature is constant.

D. The value of K will be lower at t_2 than it was at t_1.

Question 22

A reversible reaction occurs when solutions of Fe^{3+} ions are added to solutions containing thiocyanate, SCN^-, ions. The product formed is red in colour. The equation for the reaction is shown below.

$$Fe^{3+}(aq) + SCN^-(aq) \rightleftharpoons FeSCN^{2+}(aq)$$

20 mL of water is added to an equilibrium mixture of the species above. When equilibrium is re-established, the impact of this addition leads to the following possible responses.

Consider the following statements.

Statement number	Statement
1	a decrease in red intensity
2	a decrease in the value of K
3	an increase in the concentration of SCN^-
4	an increase in the amount of SCN^-

Which of the following sets of statements is correct?

A. statement numbers 1, 2 and 4

B. statement numbers 1, 3 and 4

C. statement numbers 1 and 3

D. statement numbers 1 and 4

Use the following information to answer Questions 23 and 24.

Electrolysis is conducted on a dilute solution of copper(II) sulfate, $CuSO_4$.

Copper electrodes are used. They are weighed before the circuit is switched on and after it is switched off.

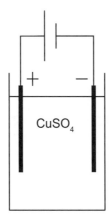

CuSO$_4$

Question 23

The reaction occurring at the anode is

A. $Cu(s) \rightarrow Cu^{2+}(aq) + 2e^-$

B. $2H_2O(l) + 2e^- \rightarrow H_2(g) + 2OH^-(aq)$

C. $2H_2O(l) \rightarrow O_2(g) + 4H^+(aq) + 4e^-$

D. $Cu^{2+}(aq) + 2e^- \rightarrow Cu(s)$

Question 24

When the circuit is closed and electrolysis is occurring, the

A. concentration of Cu^{2+} ions in solution will increase due to the anode reaction.

B. concentration of SO_4^{2-} ions in solution will drop as Cu^{2+} ions deposit on the cathode.

C. concentration of the $CuSO_4$ solution will be unchanged.

D. concentration of Cu^{2+} ions in solution will decrease due to the cathode reaction.

Question 25

Copper electrodes are placed in a dilute solution of $ZnSO_4$.

When a current is passed through the solution, which one of the following will occur?

A. Oxygen gas will be produced at the anode and zinc metal deposited at the cathode.

B. Oxygen gas will be produced at the anode and hydrogen gas at the cathode.

C. The copper anode will react and form ions and copper metal is deposited at the cathode.

D. The copper anode will react and form ions and zinc metal is deposited at the cathode.

> *Use the following information to answer Questions 26 and 27.*
>
> A white, crystalline powder is added to a crucible and heated in a fume cupboard until it melts. Inert electrodes are placed in the crucible and a current passed through the solution.
>
> A current of 10 A running for 193 s causes an increase in mass at the cathode of 0.138 g. A gas is collected above the anode. The volume of the gas after it has cooled to SLC is 0.248 L.
>
>

Question 26

The equation for the reaction occurring at the cathode in this cell is

A. $2Br^-(l) \rightarrow Br_2(g) + 2e^-$

B. $Li(l) \rightarrow Li^+(l) + e^-$

C. $Li^+(l) + e^- \rightarrow Li(l)$

D. $Na^+(l) + e^- \rightarrow Na(l)$

Question 27

What is the gas produced at the anode?

A. chlorine

B. oxygen

C. carbon dioxide

D. nitrogen

Question 28

Which one of the following electrolytic cells will produce the highest number of mole of metal?

A. a current of 9650 A running for 1000 seconds through $AlCl_3$(aq)

B. a current of 9650 A running for 100 seconds through $CuCl_2$(aq)

C. a current of 9650 A running for 90 seconds through $AgNO_3$(aq)

D. a current of 9650 A running for 1000 seconds through $AlCl_3$(l)

> *Use the following information to answer Questions 29 and 30.*
>
> Magnesium metal is produced commercially from the electrolysis of molten $MgCl_2$. This process is known as the Dow process. The magnesium chloride required is obtained from seawater.

Question 29

The half-equation occurring at the cathode in this cell is

A. Mg^{2+}(l) + 2e$^-$ → Mg(l)

B. $2Cl^-$(l) + 2e$^-$ → Cl_2(g)

C. $2Cl^-$(l) → Cl_2(g) + 2e$^-$

D. $2H_2O$(l) → O_2(g) + 4H$^+$(aq) + 4e$^-$

Question 30

A large current of 12 600 A is used in this process.

At SLC, the volume of gas, in litres, obtained from this cell in 1.0 hour will be

A. 2900

B. 5800

C. 11 700

D. 23 300

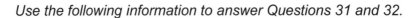

Use the following information to answer Questions 31 and 32.

Electrolysis is conducted on molten solutions of lithium, sodium and a third metal. The graph below shows the mass of each metal formed at the negative electrode as the amount of charge increases.

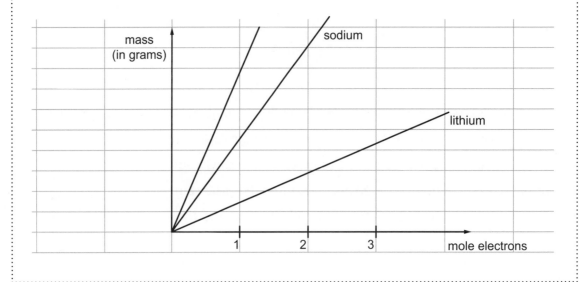

Question 31

Which of the following metals could be the third metal?

A.   silver

B.   magnesium

C.   potassium

D.   calcium

Question 32

The same data is used to plot a different graph that replaces mass on the vertical axis with number of mole of metal. On this graph

A.   all metals will produce the same line.

B.   the line for lithium will be the same as that of sodium.

C.   the line for sodium will have a lower gradient than the line for lithium.

D.   the line for sodium will still have a higher gradient than the line for lithium.

Short answer

Question 1 (7 marks)

The zinc–air fuel cell is powered by the oxidation of zinc in air. The battery has a high energy density and is relatively cheap to produce. It is referred to as 'mechanically rechargeable' because the zinc electrode can be replaced when it is fully oxidised and the discharge of the cell starts over again.

An outline of the cell is shown below. The electrode on the left is made from zinc and the electrode on the right is porous graphite. Air is able to flow through the graphite.

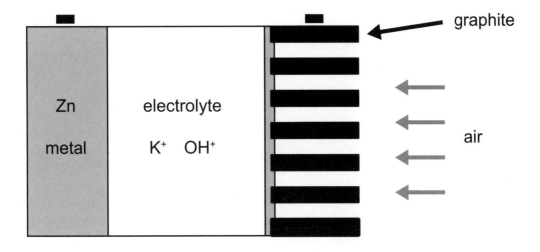

The relevant half-equations for this cell at the temperature and pH at which it will operate are

$$\tfrac{1}{2}O_2(g) + H_2O(l) + 2e^- \rightleftharpoons 2OH^-(aq) \qquad 0.34 \text{ V}$$
$$ZnO(s) + H_2O(l) + 2e^- \rightleftharpoons Zn(s) + 2OH^-(aq) \qquad -1.25 \text{ V}$$

a. Write a balanced half-equation for the reaction occurring at the   2 marks

   i.   Anode: _____

  ii.   Cathode: _____

b. **i.**   Write an overall equation for this cell.   1 mark

       _____

  ii.   What is the theoretical voltage of this cell?   1 mark

                 V

c.   i.   Give two reasons why the cell is relatively inexpensive.   2 marks

_____

_____

_____

_____

ii.   Explain why this cell is considered a type of fuel cell.   1 mark

_____

_____

Question 2 (10 marks)

One of the first commercial secondary cells was the nickel–cadmium (Ni–Cd) cell. The voltage produced in each cell is 1.2 V. Ni–Cd batteries are still popular in devices due to the high number of recharges they can sustain.

The reduction half-equation occurring during discharge is

$$NiO(OH)(s) + H_2O(l) + e^- \rightarrow Ni(OH)_2(s) + OH^-(aq)$$

The overall equation is

$$2NiO(OH)(s) + Cd(s) + 2H_2O(l) \rightarrow 2Ni(OH)_2(s) + Cd(OH)_2(s)$$

a.   i.   What is the oxidation state change of nickel ions during discharge?   1 mark

_____

ii.   Which metal is the stronger reducing agent, nickel or cadmium?   1 mark

_____

iii.   Write a balanced half-equation for the oxidation reaction occurring.   1 mark

_____

b. A set of Ni–Cd cells is in the process of being recharged, as shown below.

Positive electrode:

Negative electrode:

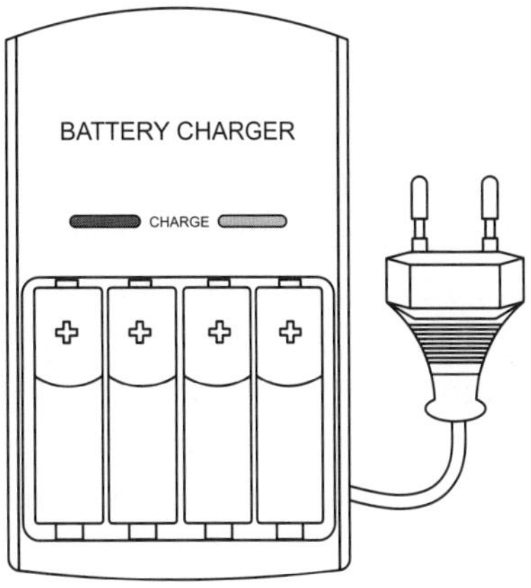

BATTERY CHARGER

CHARGE

i. Use the boxes provided above to write balanced half-equations for the reactions occurring during the recharge of this cell.   2 marks

ii. What voltage should the recharger be using in this process?   1 mark

iii. List one similarity between Ni–Cd cells and fuel cells.   1 mark

c. Methane can be used to power fuel cells. Using the spaces provided below, show the equations that occur when a methane fuel cell is operating in acidic conditions.   3 marks

Anode half-equation: _____

Cathode half-equation: _____

Overall equation: _____

Question 3 (12 marks)

Zinc–bromine flow batteries have been available for several years. These batteries use an electrolyte of zinc bromide solution, which is connected to tanks to store the reactants. The flow batteries perform well, but the need for separate tanks of liquid makes them unsuitable for many applications.

A Sydney company is trialling a modification to the design, using a gel. The gel holds reactants without the need for separate tanks, and it can be used in moving objects because it does not splash. In the trial, the zinc–bromine gel-ion batteries will power street lights at night as they discharge. During the day, a solar array is used to recharge the battery, re-forming the reactants. A representation of this arrangement is shown below. One electrode is made from zinc and the other is graphite. The electrolyte is a gel containing zinc bromide, $ZnBr_2$. The gel also stores the bromine liquid needed for the reaction.

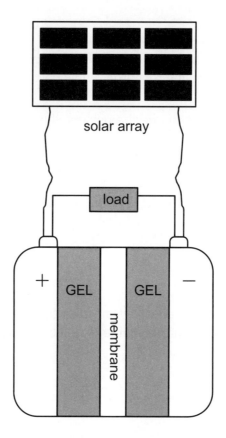

a. i. Write the half-equations occurring when the cell is discharging.

States are not required.   3 marks

	Half-equation	**Polarity**
Anode		
Cathode		

ii. Write the overall discharge equation.   1 mark

_____

© Insight Publications

iii. Write the half-equations for the recharge process. States are not required.   3 marks

	Half-equation	**Polarity**
Anode		
Cathode		

b. The voltage produced in the cell is 1.6 V, which is lower than the predicted value derived from the electrochemical series.

    i. What is the predicted voltage of this cell?   1 mark

    ii. Suggest one reason why the voltage obtained is lower than the predicted value.   1 mark

c. The mass of zinc in the electrode of each cell is 520 g. The current required to run the street light is 4.8 A.

Calculate the theoretical maximum time the light can run before a recharge is necessary.   3 marks

Question 4 (8 marks)

An experimental galvanic cell is being trialled that uses lithium metal and sulfur as reactants. The overall equation for the reaction in the cell is

$$16Li + S_8 \rightarrow 8Li_2S$$

States are not shown in this equation because a polymer electrolyte is used in the cell rather than an aqueous solution. The cell produces a voltage of 2.4 V. It is light in weight and the reactants are relatively cheap.

a.   Explain why an aqueous solution is not used in this cell. Include a balanced equation to justify your answer.   2 marks

_____

_____

_____

b.   Write a balanced half-equation for the reactions occurring at the anode and the cathode in this cell.   2 marks

Anode: _____

Cathode: _____

c.   This cell is rechargeable. Write a balanced equation for the recharging reaction in this cell.   1 mark

_____

d.   In one trial, the cell produced a current of 2.5 A for 15 minutes. Calculate the mass change at the lithium electrode during this time.   3 marks

_____

_____

_____

_____

Question 5 (6 marks)

The equation for the reaction between magnesium and hydrochloric acid is

$$Mg(s) + 2HCl(aq) \rightarrow MgCl_2(aq) + H_2(g)$$

An experiment is conducted where 1.0 g of magnesium is added to excess hydrochloric acid at 25°C and the volume of hydrogen gas evolved is monitored. The volume of gas produced is shown on the graph below.

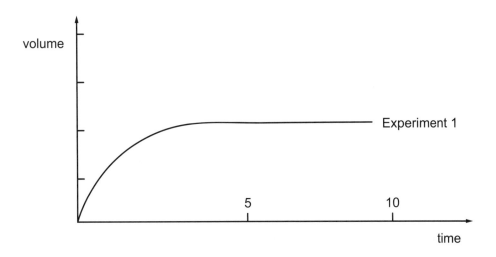

Two further experiments are conducted and the hydrogen gas evolved is shown on the graph below.

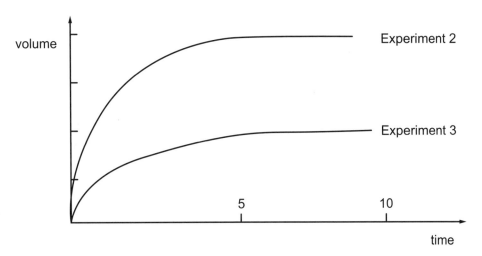

a. Suggest a change that was made to the original experiment that would have led to the hydrogen production being that shown for Experiment 2.   1 mark

b. Suggest three possible changes to Experiment 1 that might have led to the hydrogen production being that shown for Experiment 3.   3 marks

_____

_____

_____

c. List two other methods that could be used to monitor the rate of this reaction.   2 marks

_____

_____

Question 6 (7 marks)

The reaction between bromine gas and methanoic acid is

$$Br_2(aq) + HCOOH(aq) \rightarrow 2Br^-(aq) + 2H^+(aq) + CO_2(g)$$

Br_2 gas has a brown colour, but Br^- ions are colourless. The reaction is not reversible but the reaction rate is relatively slow.

a. A 1.0 mole sample of bromine is reacted with a 0.8 mole sample of methanoic acid.

i. Describe the likely change in brown intensity an observer will see.   1 mark

_____

_____

ii. The reaction is repeated with the same amount of chemicals but this time with a catalyst added. Compare the change in brown intensity of this reaction with that in the first reaction.   2 marks

_____

_____

b. **i.** The reaction is repeated in a flask that is sitting on a balance. Use the axes below to draw how the mass of the flask will change as the reaction proceeds.

1 mark

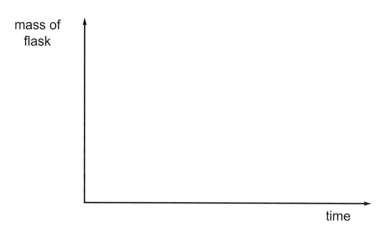

ii. Calculate the final mass loss of the flask.

2 marks

c. The graph below shows the volume of CO_2 produced in a reaction between 1.0 mole of Br_2 and 1.0 mole of methanoic acid.

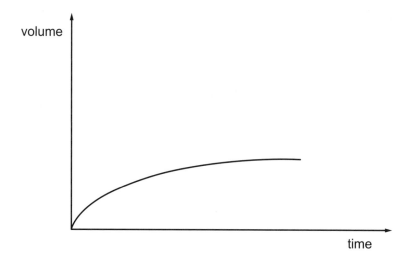

The reaction is now repeated but the amount of Br_2 is doubled. On the axes above, draw the graph of the CO_2 produced from this reaction.

1 mark

Question 7 (8 marks)

Carbon monoxide and hydrogen can be formed from the reaction of methane with steam according to the equation

$$CH_4(g) + H_2O(g) \rightleftharpoons CO(g) + 3H_2(g), \Delta H = +206 \text{ kJ}$$

a. 2.0 mole of methane and 1.8 mole of steam are added to an empty 1.0 L reactor. When equilibrium is reached, the amount of carbon monoxide formed is 0.24 mole.

Calculate the value of the equilibrium constant for this reaction.   3 marks

b. The volume of an equilibrium mixture of the abovementioned gases is halved at time t_1. The mixture returns to equilibrium at time t_2. The temperature is held constant.

At time t_2, how does the

i.   value of K compare to the value of K just before t_1?   1 mark

ii. amount of CO compare to the amount of CO just before t_1?   1 mark

iii. concentration of CO compare to the concentration of CO just before t_1?   1 mark

iv. rate of the forward reaction compare to the rate of forward reaction just before t_1?   1 mark

c. The value of the equilibrium constant for the reaction above is K. Write a balanced equation for the reaction that will have an equilibrium constant of $\frac{1}{K}$ at the same temperature.   1 mark

Question 8 (8 marks)

Methanol gas is unstable and can decompose to hydrogen and carbon monoxide in a reversible reaction.

The equation for the reaction is

$$CH_3OH(g) \rightleftharpoons CO(g) + 2H_2(g), \Delta H > 0.$$

The value of K for this reaction at 760 °C is 3.52×10^{-3} M^2.

a. A sample of methanol is added to an empty reactor at 760 °C. When equilibrium is established, the concentration of methanol is determined to be 0.84 M.

Calculate the equilibrium concentration of CO.   4 marks

b. A reactor contains an equilibrium mixture of the aforementioned gases. The volume of the reactor is halved.

i. Explain the impact of this change on the position of equilibrium of this reaction.   2 marks

ii. How will the concentration of CO after equilibrium is re-established compare with its value before the volume was changed?   1 mark

iii. How will the rate of reaction in the system after equilibrium is re-established compare with the rate before the volume was changed?   1 mark

Question 9 (10 marks)

The equation for the formation of sulfur trioxide from sulfur dioxide and oxygen gases is

$$2SO_2(g) + O_2(g) \rightleftharpoons 2SO_3(g), \qquad \Delta H = -ve$$

During an experiment, equal amounts of SO_2 and O_2 gases are added to an empty reactor at 100 °C. The graph below shows how the concentration of O_2 changes as the system moves to equilibrium.

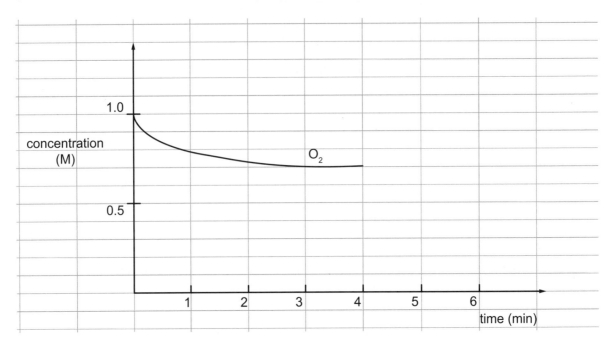

a. **i.** Using the axes provided above, draw the concentrations of both SO_2 and SO_3 gases for the first 4 minutes of the experiment.   2 marks

   ii. Determine the value of K at 100 °C for this reaction.   2 marks

b. At the 4-minute mark, a sample of oxygen is injected into the reactor, increasing the immediate concentration of O_2 to 0.8 M.

Show this increase on the axes above and draw how the concentrations of all three species will respond as the system moves to re-establish equilibrium.   3 marks

c. The temperature of an equilibrium mixture of the three gases is increased from 100 °C to 200 °C.

Identify the impact of this temperature change on the

   i. value of K.   1 mark

ii.   amount of SO$_3$ gas.   1 mark

_____

iii.   total amount of gas.   1 mark

_____

Question 10 (10 marks)

A sample of sodium chloride is added to a crucible and heated to a high temperature. When the solid melts, two carbon electrodes are placed in the liquid and connected to a power supply. The power supply is switched on.

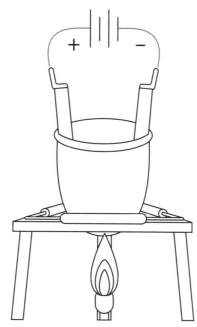

a.   Describe the movement of the particles in the liquid.   2 marks

_____

_____

b.   Write half-equations for the reactions occurring at the   2 marks

i.   anode: _____

ii.   cathode: _____

c.   Write a balanced equation for the overall reaction that is occurring.   1 mark

_____

d. Give two reasons why this experiment should not be conducted in a school laboratory.   2 marks

_____

_____

e. If the current is 5.62 A, determine the time required for 0.460 mole of gas to be produced in the cell.   3 marks

_____

_____

_____

_____

_____

_____

Question 11 (9 marks)

Calcium can be manufactured by the reduction of lime (CaO), using aluminium metal. An alternative process is where a molten mixture of calcium fluoride and calcium chloride is electrolysed. The electricity costs of this method are high but the purity of the calcium obtained is superior to that obtained by other methods. An outline of the cell used is shown below.

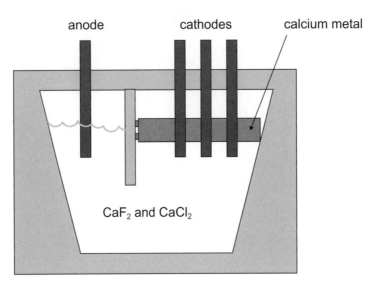

a. Write a balanced equation for the reduction of CaO by aluminium metal.   1 mark

_____

b. The electrolyte used in the cell shown on the previous page is a mixture of CaF_2 and $CaCl_2$.

Write balanced half-equations for the

　i. reaction at the anode.　　　　　　　　　　　　　　　　　　　1 mark

　ii. reaction at the cathode.　　　　　　　　　　　　　　　　　1 mark

　iii. overall reaction.　　　　　　　　　　　　　　　　　　　　1 mark

c. The current in the cell is 125 000 A.

How long will it take (in hours) to obtain 1.00 tonne of calcium metal?　　3 marks

d. Potassium salts are often found in the same deposits as calcium salts.

Use your knowledge of the electrochemical series to explain whether the presence of potassium salts is likely to interfere with the operation of the calcium cell.　　　　　　　　　　　　　　　　　　　　　　2 marks

Question 12 (9 marks)

Graphite electrodes are placed into a dilute nickel(II) chloride solution and a power supply is connected.

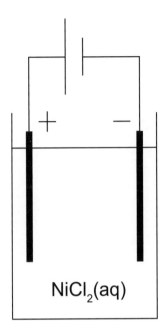

NiCl$_2$(aq)

a. Use the electrochemical series to predict the reactions that will occur in this cell when the power supply is switched on.   4 marks

Anode: _____

Cathode: _____

Overall equation: _____

b. Describe what will be observed happening at each electrode.   2 marks

Anode: _____

Cathode: _____

c. The circuit operates until 2.50 g of nickel has been deposited. Calculate the volume of gas produced if the cell is running at SLC conditions.   3 marks

_____

_____

_____

_____

_____

Question 13 (11 marks)

Various forms of electrolyser are used to produce green hydrogen.

a. i.   What is green hydrogen?   1 mark

ii.   State two important uses of hydrogen gas.   2 marks

b.   A dilute solution of sulfuric acid is often electrolysed in school laboratories to produce hydrogen gas. Write the half-equations for this process.   2 marks

Anode: _____

Cathode: _____

c.   The diagram below is a sketch of a PEM (polymer electrolytic membrane) cell that operates in acidic conditions.

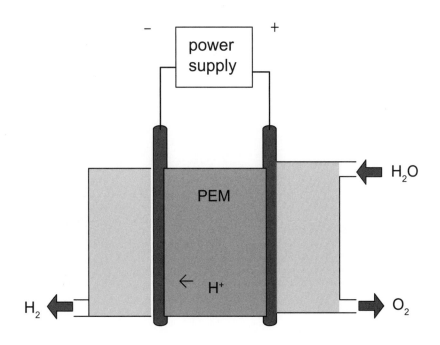

Write half-equations for the reactions occurring in a PEM cell (states not required).   2 marks

Anode: _____

Cathode: _____

d.   The use of hydrogen as a fuel offers significant potential to reduce our current dependence on fossil fuels. However, there are still limitations to overcome before hydrogen is accepted for wide-scale commercial use.

Identify two limitations of hydrogen gas at this time and explain why each is a limitation.   4 marks

Limitation 1: _____

_____

Limitation 2: _____

_____

Unit 4 | Area of Study 1   How are organic compounds categorised and synthesised?

Multiple choice

Question 1

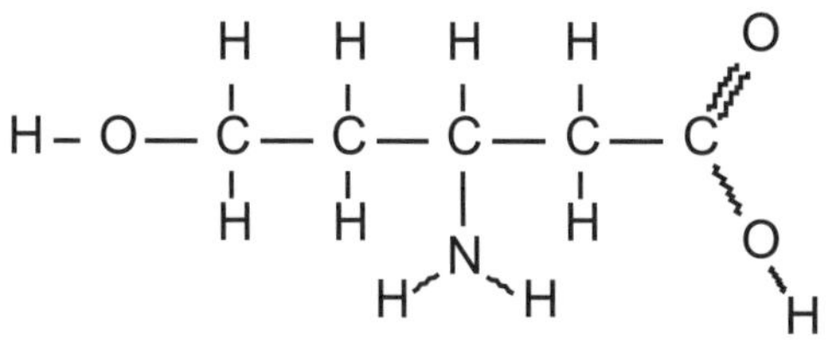

The functional groups present on this molecule are

A. hydroxyl, amide and carboxyl.

B. hydroxyl, amine and carboxyl.

C. hydroxyl, amide and ester.

D. hydroxyl, amine and carbonyl.

Question 2

The general formula for a linear alcohol is

A. $C_nH_{2n}O$

B. $C_nH_{2n+1}O$

C. $C_nH_{2n+2}O$

D. $C_nH_{2n}O_2$

Question 3

Consider the following molecules.

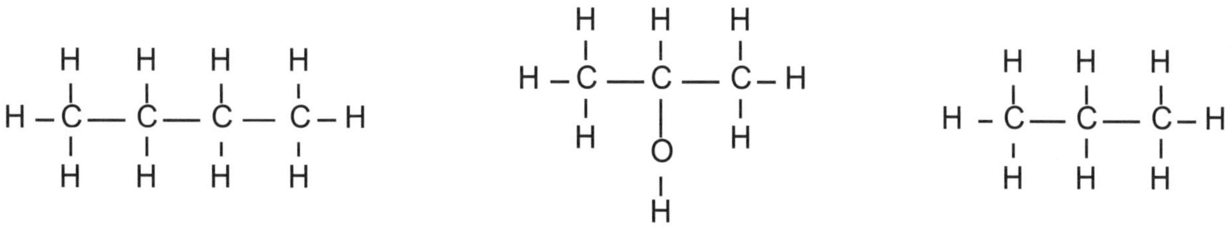

The boiling point, from lowest to highest, for these molecules is

A. propane, butane, propan-2-ol.

B. propane, propan-2-ol, butane.

C. butane, propan-2-ol, propane.

D. propan-2-ol, propane, butane.

Question 4

The systematic name for the molecule shown above is

A.   3-methylbutan-1-ol.

B.   2-hydroxybutane.

C.   3-methylbutan-4-ol.

D.   2-methylbutan-1-ol.

Question 5

Consider the following molecules.

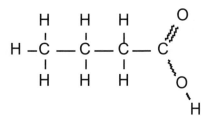

The systematic names for the molecules shown are, from left to right

A.   propan-2-ol, prop-2-ene and 1-butanoic acid.

B.   propan-2-ol, propene and butanoic acid.

C.   propan-2-ol, propene and 1-butanoic acid.

D.   propan-1-ol, propene and butanoic acid.

Question 6

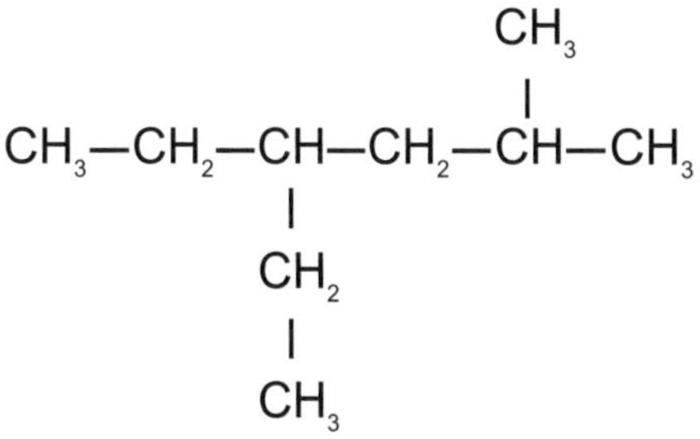

$$CH_3-CH_2-CH-CH_2-\underset{\underset{\underset{CH_3}{|}}{\underset{CH_2}{|}}{\overset{\overset{CH_3}{|}}{CH}}}{\underset{|}{CH}}-CH_3$$

The IUPAC name for the molecule shown above is

A.   4-ethyl-2-methylhexane.

B.   3-ethyl-5-methylhexane.

C.   2-methyl-5,5-diethylbutane.

D.   nonane.

Question 7

A short organic pathway is drawn below.

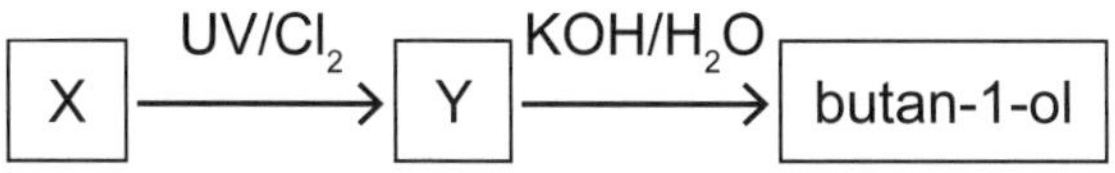

$$\boxed{X} \xrightarrow{UV/Cl_2} \boxed{Y} \xrightarrow{KOH/H_2O} \boxed{butan\text{-}1\text{-}ol}$$

In this process, molecule Y could be

A.   butene.

B.   butane.

C.   butanol.

D.   1-chlorobutane.

Question 8

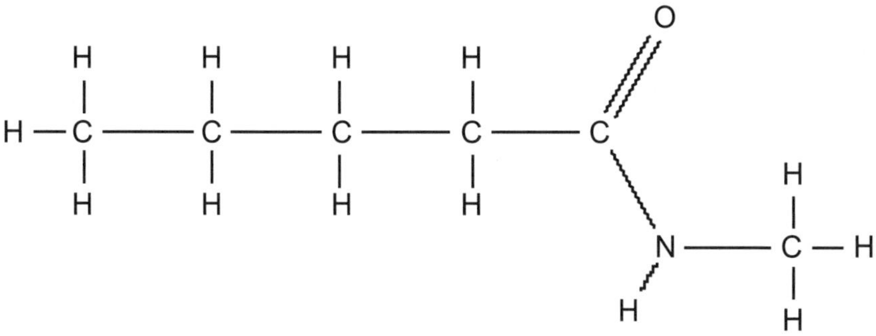

The molecule shown could be formed from a reaction between

A.   methanoic acid and pentanamine.

B.   methanamine and pentanoic acid.

C.   ethanamine and pentanoic acid.

D.   ethanamine and pentan-1-ol.

Question 9

The reaction of molecules A and B forms the products shown below.

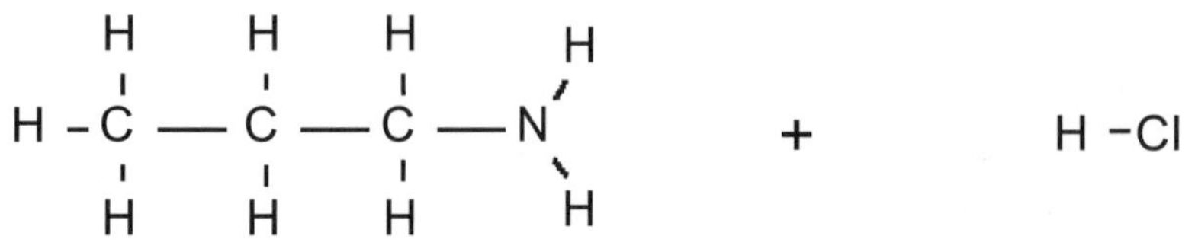

$$H-\underset{\underset{H}{|}}{\overset{\overset{H}{|}}{C}}-\underset{\underset{H}{|}}{\overset{\overset{H}{|}}{C}}-\underset{\underset{H}{|}}{\overset{\overset{H}{|}}{C}}-N\overset{H}{\underset{H}{\diagdown}}\qquad+\qquad H-Cl$$

Molecules A and B could be

A. methane and chloroethane.

B. propan-1-ol and methanamine.

C. 2-chloropropane and methanamine.

D. 1-chloropropane and ammonia, NH_3.

Question 10

The semi-structural formula of an ester is $CH_3CH_2OCOCH_3$.

This molecule could be formed from a reaction between

A. ethanol and methanoic acid.

B. ethanoic acid and methanol.

C. propan-1-ol and methanoic acid.

D. ethanoic acid and ethanol.

Question 11

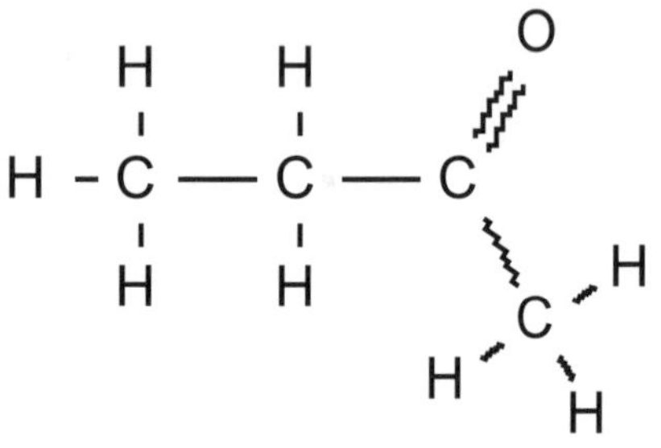

$$H-\underset{\underset{H}{|}}{\overset{\overset{H}{|}}{C}}-\underset{\underset{H}{|}}{\overset{\overset{H}{|}}{C}}-C\overset{\displaystyle O}{\underset{\underset{\underset{H}{\diagup}\overset{|}{\underset{H}{\diagdown}}H}{C}}{\diagdown}}$$

The molecule shown above could be formed from

A. 2-chlorobutane, using a substitution reaction followed by an oxidation reaction.

B. 2-chlorobutane, using a substitution reaction followed by a condensation reaction.

C. 1-chlorobutane, using an addition reaction followed by an oxidation reaction.

D. a condensation reaction between methanol and propanoic acid.

Short answer

Question 1 (7 marks)

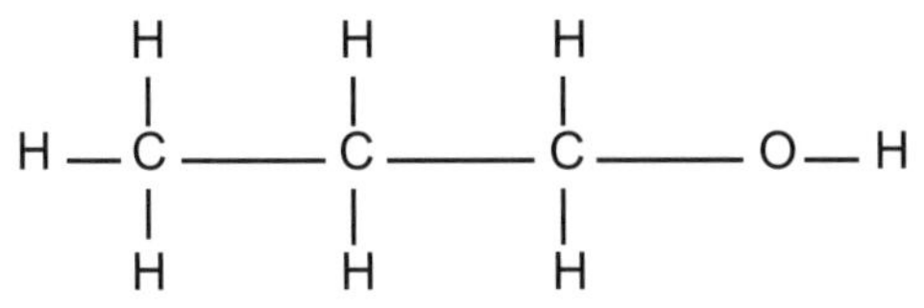

propanol

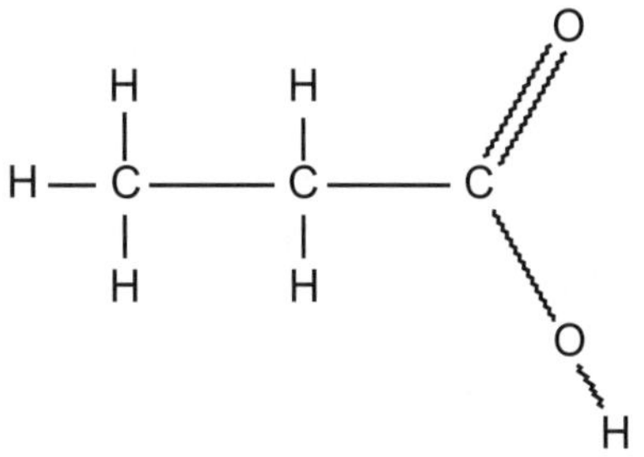

1-propanoic acid

a. The names provided for these molecules are incorrect IUPAC names.

Explain why.   2 marks

propanol: _____

_____

1-propanoic acid: _____

_____

b. **i.**

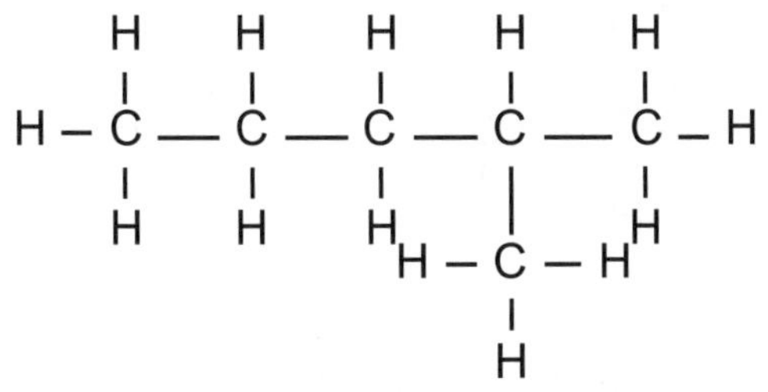

4-methyl pentane is not the correct systematic name for this compound.

Explain why.   1 mark

_____

ii. Is this molecule an isomer of pentane?   1 mark

_____

c.   **i.**   Name the following two molecules.   2 marks

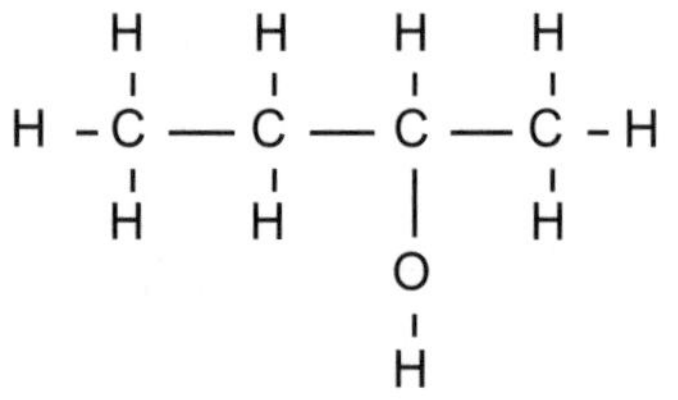

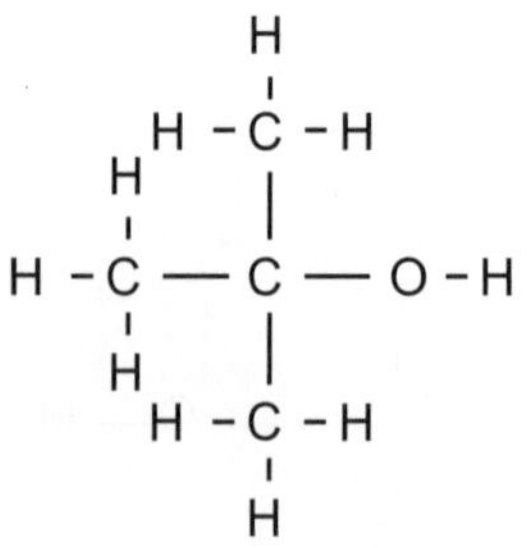

_____   _____

ii.   Use the two molecules shown in **part c.i.** to explain the difference between a
secondary alcohol and a tertiary alcohol.   3 marks

_____

_____

Question 2 (8 marks)

a.   The molecules of ethane and hexane, respectively, are shown below.

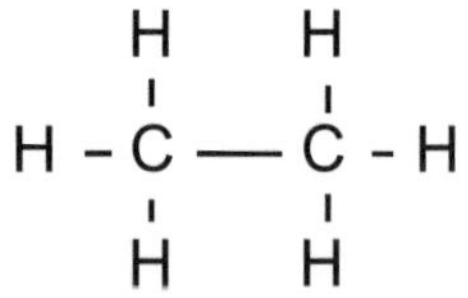

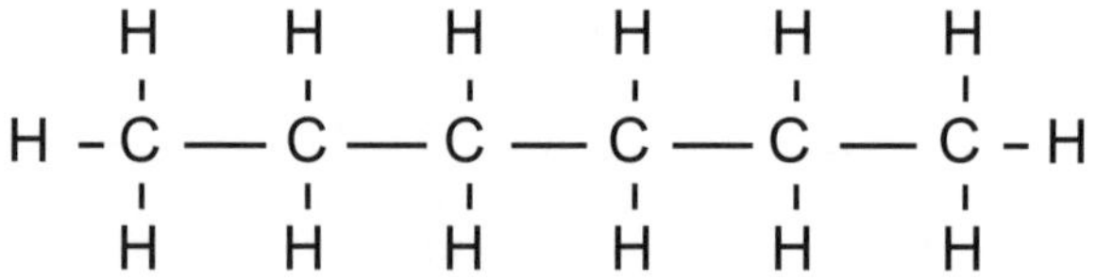

i.   State a property of the two molecules that is similar and give a reason
for the similarity.   2 marks

_____

_____

ii.   State a property of the two molecules that is different and give a reason
for the difference.   2 marks

_____

_____

b. Ethyl propanoate is an ester. There are several accepted ways of representing the structure of this molecule. Give a representation for each of the four ways listed below.

i. molecular formula: _____ 1 mark

ii. structural diagram: 1 mark

iii. semi-structural formula: _____ 1 mark

iv. skeletal structure: 1 mark

Question 3 (9 marks)

The molecule shown below is an ester.

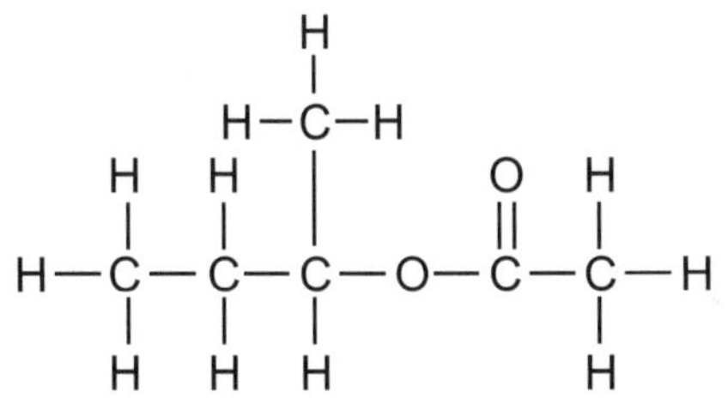

a. **i.** Write the molecular formula of this ester. 1 mark

_____

ii. Write its empirical formula. 1 mark

_____

The molecule is hydrolysed using NaOH solution to form an alcohol and an ionic salt.
The products are separated and HCl is added to the salt to convert it to a carboxylic acid.
This is shown in the diagram on the next page.

b.

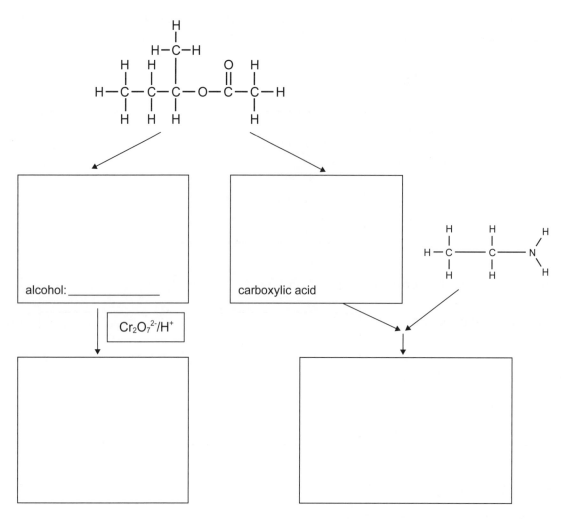

i.   In the relevant box provided above, draw the structural formula of the
     alcohol molecule formed.   1 mark

ii.

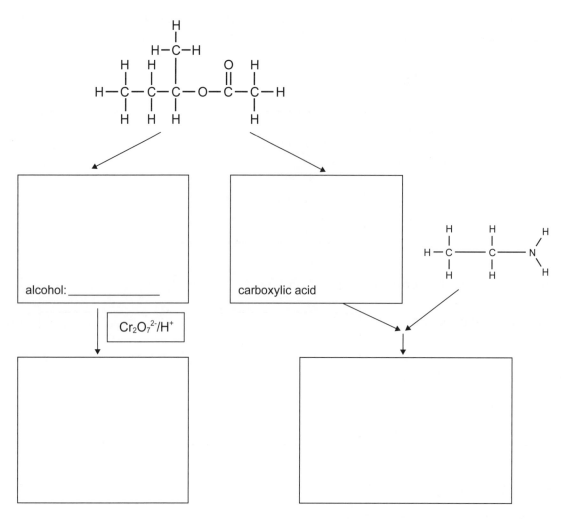

The molecule above contains a chiral centre. Mark this centre with
a star.   1 mark

iii. On the line provided in the box, write the name of the alcohol.   1 mark

iv. The alcohol is reacted with acidified $Cr_2O_7^{2-}$ solution. In the relevant box
     provided above draw the structural formula of the molecule that will form.   1 mark

c. i.   In the box provided, draw the structural formula of the carboxylic
             acid formed when the ester is hydrolysed.   1 mark

ii. The carboxylic acid is reacted with the molecule shown as molecule M
     in **part b.i.** In the box provided, draw the structural formulas of the
     products formed.   2 marks

Question 4 (8 marks)

Ethanol is an important chemical, with world production exceeding 112 billion litres in 2023. A variety of production processes are used to meet this demand for ethanol, some of which are outlined below.

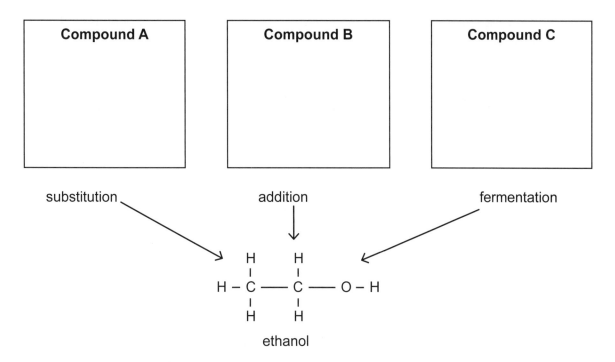

a. Ethanol can be produced through substitution reactions.

    i. In the box labelled Compound A, draw an example of a molecule that could be used to manufacture ethanol by substitution.   1 mark

    ii. Write a balanced equation for the formation of ethanol from Compound A. States are not required.   1 mark

_____

b. Ethanol can be produced through addition reactions.

    i. In the box labelled Compound B, draw an example of a molecule that could be used to manufacture ethanol by addition.   1 mark

    ii. Write a balanced equation for the formation of ethanol from Compound B. States are not required.   1 mark

_____

c.   Ethanol can be produced by fermentation.

   i.   In the box labelled Compound C, draw an example of a molecule that
          could be used to manufacture ethanol by fermentation.   1 mark

   ii.   Write a balanced equation for the formation of ethanol from Compound C.
          States are not required.   1 mark

d.   Identify which of the processes for manufacturing ethanol given above will have
       the highest atom economy.

       Justify your answer.   2 marks

Question 5 (9 marks)

The flowchart below can be followed to produce the amide molecule shown. In the final step of the flowchart, a condensation reaction occurs between Molecules C and D.

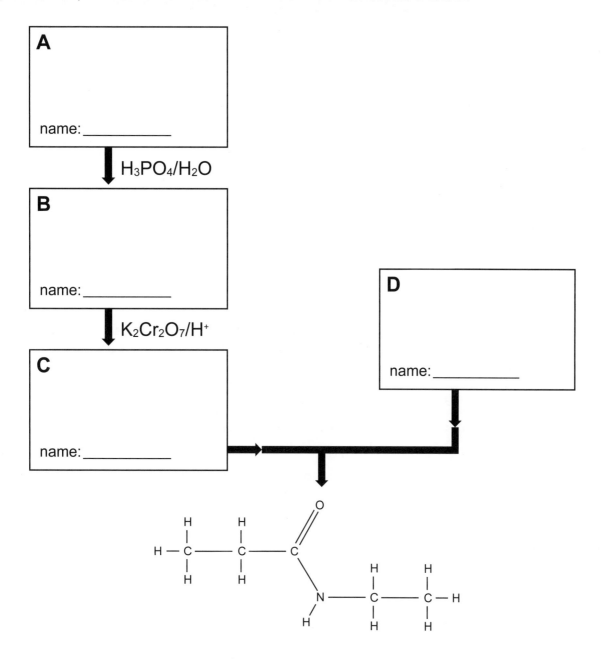

a. In the boxes provided above, draw and name Molecules A, B, C and D.   8 marks

b. In the production of Molecule C, dichromate ions are reduced to Cr^{3+} ions.

Write a balanced half-quation for this reaction.   1 mark

Unit 4 | Area of Study 2   How are organic compounds analysed and used?

Multiple choice

Question 1

An organic molecule is tested and found to turn bromine colourless and to react with NaOH. The molecule did not react with acidified $Cr_2O_7^{2-}$ ions.

The molecule could be

A. CH_2CHCH_2COOH

B. $CH_3CH_2CH_2COOH$

C. $CH_3CH_2CH_2CH_2OH$

D. $CH_2CHCH_2CH_2OH$

Question 2

A class is split into four groups. Each group is provided with a solution of ethanoic acid and asked to determine the concentration of the acid using a titration. The results obtained by each group are shown below.

	Group 1	Group 2	Group 3	Group 4
Concentration obtained	0.123 M	0.122 M	0.122 M	0.124 M

The concentration of the ethanoic acid is tested in a commercial laboratory and found to be 0.186 M.

The results of the class can be described as

A.   neither precise nor accurate.

B.   precise but not accurate.

C.   accurate but not precise.

D.   accurate and precise.

Use the following information to answer Questions 3 and 4.

A titration is conducted to determine the concentration of a solution of iron(II) sulfate.

The standard used is a solution of 0.260 M potassium permanganate. The reaction can be represented by the following equation.

$$5Fe^{2+}(aq) + MnO_4^-(aq) + 8H^+(aq) \rightarrow 5Fe^{3+}(aq) + Mn^{2+}(aq) + 4H_2O(l)$$

Question 3

A burette is rinsed with deionised water before it is filled with potassium permanganate solution.

The use of water for rinsing

A.   is good practice.

B.   will lead to a low titre and a high estimate of iron concentration.

C.   will lead to a low titre and a low estimate of iron concentration.

D.   will lead to a high titre and a high estimate of iron concentration.

Question 4

20.0 mL aliquots of potassium permanganate solution are used and the mean titre of iron solution is 25.0 mL.

The concentration of the iron solution, in M, is

A.   1.04

B.   0.42

C.   0.21

D.   0.042

Question 5

An organic compound is found to react with acidified dichromate ions, $Cr_2O_7^{2-}$. The product formed has a pH of 3. The organic compound could be

A.   propan-1-ol.

B.   propan-2-ol.

C.   propanoic acid.

D.   propanone.

Question 6

A chemist investigates the components of a mixture obtained from the bark of a tree, hoping to isolate a sample of salicylic acid. The documented melting point of salicylic acid is 159 °C. The melting point of the sample extracted is measured as being between 165 and 170 °C. A valid conclusion is that the extract

A.   is very high-purity salicylic acid.

B.   is unlikely to be salicylic acid.

C.   is salicylic acid with a low level of impurity.

D.   contains traces of salicylic acid.

Question 7

The molecular formula of a food molecule is $C_{65}H_{102}N_{18}O_{21}$.

The molecule could be

A. collagen.

B. triglyceride formed from stearic acid and glycerol.

C. polysaccharide starch.

D. polysaccharide cellulose.

Question 8

How many different tripeptides can be formed from reaction of the amino acids leucine, lysine and isoleucine?

A. 1

B. 3

C. 4

D. 6

Question 9

Hydrolysis of a particular macronutrient produces many carboxyl groups and hydroxyl (alcohol) groups.

The macronutrient is most likely to be a

A. triglyceride.

B. protein.

C. carbohydrate.

D. vitamin.

Question 10

Which of the following best describes the role of lipase in digestion?

A. Lipase catalyses the hydrolysis of ester bonds in foods.

B. Lipase acts to transport fatty acid molecules around the body.

C. Lipase hydrolyses the amide bonds in proteins to produce smaller amino acids.

D. Lipase acts to break large globules of fat into much smaller, more manageable globules.

Question 11

A fatty acid has the empirical formula $C_9H_{16}O$.

How many carbon-to-carbon double bonds will the fatty acid contain?

A. 0

B. 1

C. 2

D. 3

Question 12

Infrared spectroscopy is used to detect the presence of propanoic acid in a sample of propanal.

The best evidence of the presence of propanoic acid is

A. an absorption band around 1700 cm^{-1}.

B. a sharp absorption band around 3000 cm^{-1}.

C. a broad peak around 3300 cm^{-1}.

D. a broad peak around 3000 cm^{-1}.

Question 13

The high resolution proton NMR spectrum below is of 2-chloropropane.

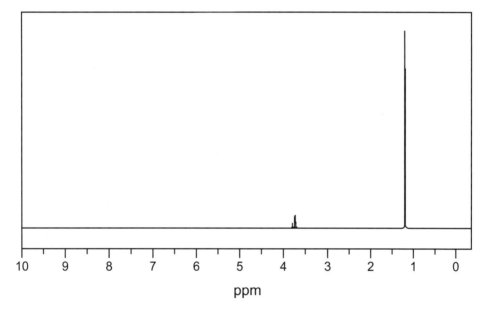

ppm

Data: SDBSWeb; http://sdbs.db.aist.go.jp, National Institute of Advanced Industrial Science and Technology

The splitting on the peak with a shift of 3.8 is not very distinct.

Given the structure of 2-chloropropane, this peak should be a

A. singlet (1 peak only).

B. doublet (split into 2).

C. sextet (split into 6).

D. septet (split into 7).

Question 14

A ^{13}C NMR spectrum is shown below.

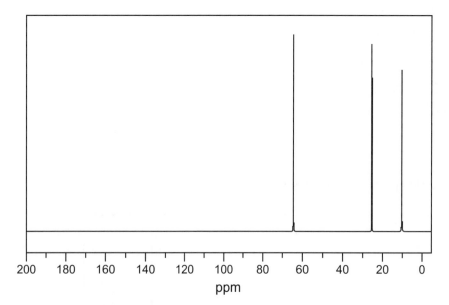

Data: SDBSWeb; http://sdbs.db.aist.go.jp National Institute of Advanced Industrial Science and Technology

The ^{13}C NMR spectrum is that of

A. propan-2-ol.

B. propanone.

C. propanoic acid.

D. propan-1-ol.

Question 15

A ^{13}C NMR spectrum is shown below.

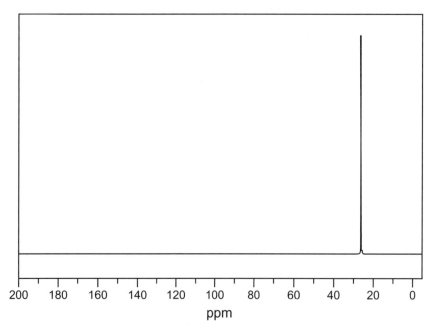

Data: SDBS Web, <http://sdbs.db.aist.go.jp>, National Institute of Advanced Industrial Science and Technology

The ^{13}C NMR spectrum could be that of

A. propane.

B. butane.

C. cyclohexene.

D. cyclohexane.

Question 16

A molecule of propane is drawn below.

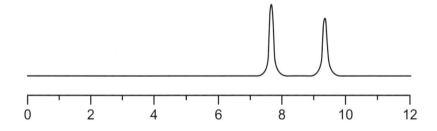

A proton NMR spectrum of propane will show

A. one singlet because all protons have the same environment.

B. two sets of peaks, one being a quartet and the other a triplet.

C. two sets of peaks, one being a septet and the other a triplet.

D. three sets of peaks, because there are three different proton environments.

Question 17

A mixture of propan-1-ol and butan-1-ol is injected into a high-performance liquid chromatography (HPLC) column that is using a polar solvent.

A second sample is passed through the same column and produces only one peak with a retention time of 9.4 minutes and an area of 1200 units. The print-out below is obtained.

Peak (min)	Area
7.6	4200
9.4	3600

The second sample contains

A. propan-1-ol and butan-1-ol at lower concentrations than the original solution.

B. propan-1-ol only, with a concentration that is one-third of the original solution.

C. butan-1-ol only, with a concentration that is one-third of the original solution.

D. butan-1-ol only, with a concentration that is three times that of the original solution.

Question 18

A sample of petrol is injected into a HPLC instrument. The stationary phase used is polar and the mobile phase is non-polar.

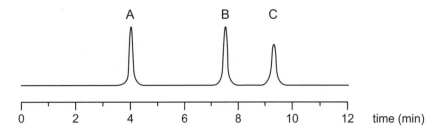

Select the correct conclusion from the following options.

A. There are only three components in the sample.

B. Molecule A is likely to be the hydrocarbon of lowest molecular mass.

C. Molecule C is likely to be the hydrocarbon of lowest molecular mass.

D. The concentration of each component is similar.

Use the following information to answer Questions 19 and 20.

A mixture containing the three molecules below is injected into a HPLC column that uses a polar stationary phase and hexane as the mobile phase.

Alcohol 1   Alcohol 2   Alcohol 3

The chromatogram obtained is shown first, and a chromatogram of a related sample is shown second. (Assume that the scales of both graphs are the same.)

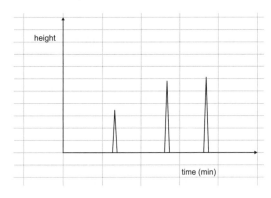

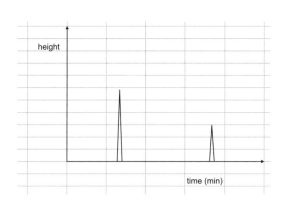

Run 1: mixture of three alcohols   Run 2: related mixture

Question 19

It is likely that Alcohol 3 will have the

A. shortest retention time because it is the most polar of the molecules shown.

B. longest retention time because it will adsorb more readily to the stationary phase.

C. shortest retention time because it has the highest relative molecular mass.

D. shortest retention time because it is the most soluble in the mobile phase.

Question 20

In Run 2, it is likely that

A. the temperature has changed, affecting the retention times of the alcohols.

B. the alcohols are the same, but the solvent has been changed to a polar solvent.

C. two alcohols are present, but the stationary phase is now non-polar.

D. there are two alcohols, one twice as concentrated as Run 1 and the other half as concentrated.

Use the following information to answer Questions 21 and 22.

Invertase is an enzyme responsible for the hydrolysis of sucrose. A common experiment is for students to add invertase to sucrose solutions at different temperatures. Benedict's solution can be added to help compare the rate of reaction at each temperature. As hydrolysis progresses, Benedict's solution turns from blue to a reddish brown.

The graph below is of the time taken for a colour change to be evident for solutions of each temperature.

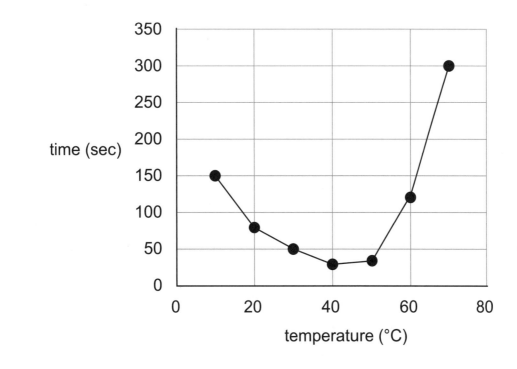

Question 21

The rate of reaction is highest at

A.   10 °C

B.   40 °C

C.   50 °C

D.   70 °C

Question 22

Which one of the following is the most likely trigger for the colour change?

A.   a reaction occurring between Benedict's solution and glucose

B.   a reaction occurring between Benedict's solution and sucrose

C.   a reaction occurring between Benedict's solution and invertase

D.   the Benedict's solution denaturing when the reaction mixtures are heated

Question 23

A piece of bread sitting in someone's mouth will taste sweet after a few minutes.

The best explanation for this effect is that

A.   the action of chewing has broken starch molecules into smaller sugar molecules.

B.   amylase in saliva has hydrolysed some starch molecules.

C.   pepsin in saliva has hydrolysed some protein molecules.

D.   the action of chewing has allowed sugar molecules to reach the taste buds.

Question 24

The diagram below is of a small snippet from a protein.

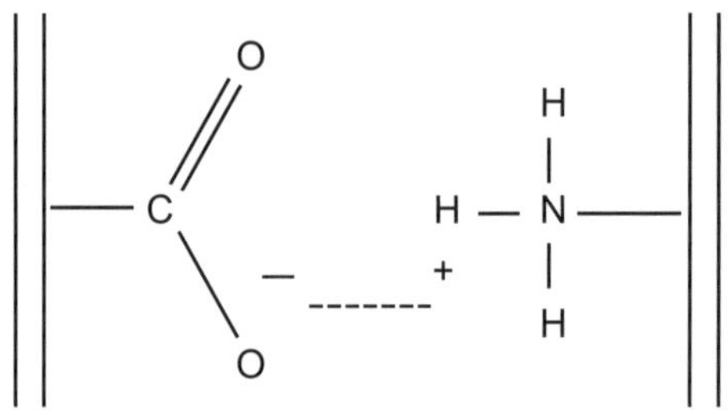

The bond in the diagram with a dashed line is an example of

A.   hydrogen bonding between different parts of a protein molecule.

B.   the hydrogen bonding that causes a protein to have a tertiary structure.

C.   an ionic bond that can form between side chains on a protein molecule.

D.   an ionic bond that forms part of the primary structure of a protein.

Question 25

The equation for the decomposition of hydrogen peroxide is $2H_2O_2(l) \rightarrow 2H_2O(l) + O_2(g)$.

This is a relatively slow reaction that can be catalysed by the addition of manganese dioxide or by the addition of pieces of liver from animals. There is an enzyme in liver that catalyses the reaction.

A 1.0 M solution of hydrogen peroxide is prepared for testing with the two catalysts.

Which alternative below will lead to one of the catalysts being very ineffective?

A.   boiling a sample of crushed liver and then adding it to 100 mL of hydrogen peroxide

B.   grinding the MnO_2 to a fine powder before adding it to 100 mL of hydrogen peroxide

C.   heating 100 mL of hydrogen peroxide to 90 °C before adding the MnO_2

D.   mincing a liver before adding it to 100 mL of hydrogen peroxide

Short answer

Question 1 (7 marks)

Oxalic acid, $C_2O_4H_2$, is sold as a white powder. It is used to bleach surfaces and as a stain remover.

A student conducts a titration to determine the percentage by mass of oxalic acid in a commercial sample of oxalic acid. He prepares a solution of oxalic acid by weighing a sample from the commercial product and adding it to a 250.0 mL volumetric flask. The volumetric flask is made up to the mark with deionised water and titrated against a 0.120 M $KMnO_4$ solution. The colour change of the permanganate ion can be used as an indicator.

The half-equation for the oxidation of oxalic acid is

$$C_2O_4H_2(aq) \rightarrow 2CO_2(g) + 2H^+(aq) + 2e^-$$

The student's measurements are recorded below.

mass of commercial oxalic acid	10.0 g
mean titre of oxalic acid	14.8 mL
MnO_4^- aliquots	20.00 mL

a.   Write a balanced equation for the reaction between oxalic acid and MnO_4^-.   2 marks

_____

b.   **i.**   Determine the number of moles of MnO_4^- in the aliquots.   1 mark

_____

_____

   ii.   Determine the number of moles of oxalic acid in the titre.   1 mark

_____

_____

   iii.   Determine the mass of oxalic acid in the volumetric flask.   2 marks

_____

_____

_____

   iv.   Calculate the % (*m/m*) of oxalic acid in the commercial product.   1 mark

_____

Question 2 (6 marks)

A number of laboratory tests are available to chemists to help analyse organic compounds. The table below gives three such tests.

Complete the table by filling in your expected observations in column 2 and brief explanations in column 3.

Test	Observation	Explanation
Acidified $Cr_2O_7^{2-}$ is heated with ethanol.		
Acidified $Cr_2O_7^{2-}$ is heated with 2-methylpropan-2-ol.		
Iodine solution is added from a burette to linolenic acid.		

Question 3 (6 marks)

Hydrolysis of macronutrients is an important step in the digestion of most macronutrients.

Complete the table below to summarise the bonds broken and the products formed during hydrolysis.

Macronutrient	Draw the bond broken during hydrolysis	Name of product(s) of hydrolysis
protein		
carbohydrate		
triglyceride		

Question 4 (4 marks)

The structure of the main component of a sample of biodiesel is shown below.

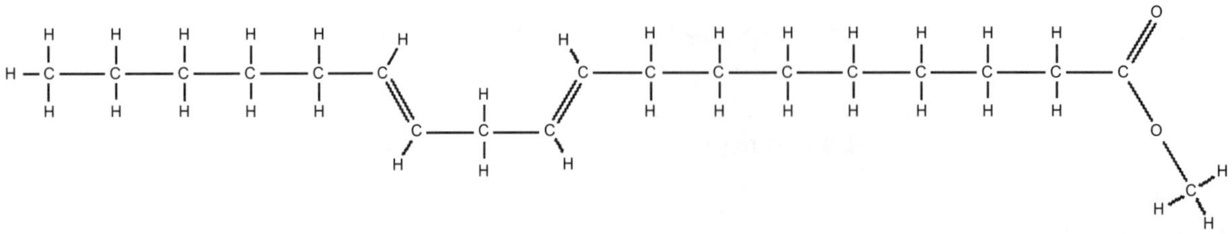

a.   This ester was formed in a condensation reaction between two smaller molecules: a fatty acid and an alcohol.

   i.   Draw the semi-structural formula of the fatty acid molecule.   1 mark

   ii.   Name the fatty acid.   1 mark

   _____

b.   Draw the semi-structural formula of the triglyceride that could be formed from a sample of this fatty acid.   2 marks

Question 5 (7 marks)

The molecule shown below is an example of a triglyceride.

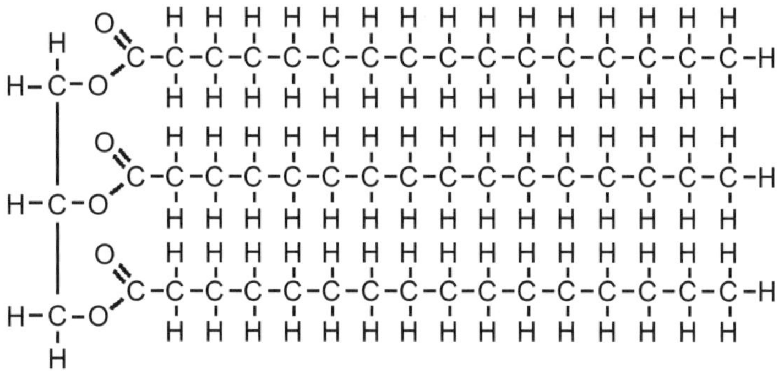

a.   i.   Triglycerides are hydrolysed in the body as part of the digestion process.

Draw structural diagrams of the two products resulting from the hydrolysis of this triglyceride.   2 marks

ii.   In which part of the digestive system is hydrolysis of a triglyceride most likely to occur? Explain your answer.   2 marks

b.   One of the products of hydrolysis of this triglyceride could be reacted with methanol to form a molecule of biodiesel.

i.   Write the semi-structural formula of the biodiesel molecule formed.   1 mark

ii.   Write a balanced equation for the complete combustion of the biodiesel molecule.   2 marks

Question 6 (6 marks)

The molecular formulas of several molecules are shown below.

$$C_{18}H_{36}O_2 \qquad\qquad C_{19}H_{38}O_2 \qquad\qquad C_6H_{12}O_6$$
A $\qquad\qquad$ **B** $\qquad\qquad$ **C**

$$C_{12}H_{22}O_{11} \qquad\qquad C_3H_7O_2NS \qquad\qquad C_3H_8O_3$$
D $\qquad\qquad$ **E** $\qquad\qquad$ **F**

$$C_{18}H_{34}O_2 \qquad\qquad C_3H_6O_2$$
G $\qquad\qquad$ **H**

Select from the above molecules to answer the following questions. A molecule can be selected more than once.

Which of the molecules **A** to **H** represents

a.   a saturated biodiesel molecule? $\hfill$ 1 mark

b.   a disaccharide? $\hfill$ 1 mark

c.   a by-product of the hydrolysis of lipids (a first step in the production of biodiesel)? $\hfill$ 1 mark

d.   an unsaturated fatty acid? $\hfill$ 1 mark

e.   a molecule that is produced using energy from the Sun? $\hfill$ 1 mark

f.   a molecule that could form a peptide bond? $\hfill$ 1 mark

Question 7 (5 marks)

The following is an example of a balanced chemical equation.

$$Mg(s) + 2HCl(aq) \rightarrow MgCl_2(aq) + H_2(g)$$

The state of each reactant and product is usually shown in an equation to convey a more accurate picture of the reaction occurring. Some reactions will not occur if the reactants are not present in the correct state.

Complete the following table by writing the state that each chemical should be in for the reaction to occur. Choose from the following four states: (s), (l), (g) or (aq).

Chemical	State
glucose in fermentation	
sulfuric acid as an esterification catalyst	
H_2O reacting with ethene to form ethanol	
biodiesel in combustion	
potassium metal produced in electrolysis	

Question 8 (10 marks)

A liquid found in a bottle has a molecular formula of C_3H_6O. There are several possible structures that match this formula.

a. Draw the structures of three molecules that have a molecular formula of C_3H_6O.

On the lines provided, write the systematic name of each molecule you have drawn. 3 marks

1. _____ 2. _____ 3. _____

The mass spectrum of the molecule is shown below.

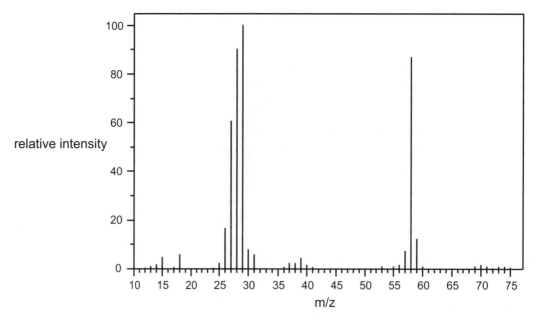

Data: SDBSWeb; http://sdbs.db.aist.go.jp, National Institute of Advanced Industrial Science and Technology

b. **i.** What is the m/z ratio of the base peak for this molecule?   1 mark

_____

ii. Suggest two possible fragments that could cause this peak.   2 marks

_____

_____

c. The infrared spectrum of the molecule is shown below.

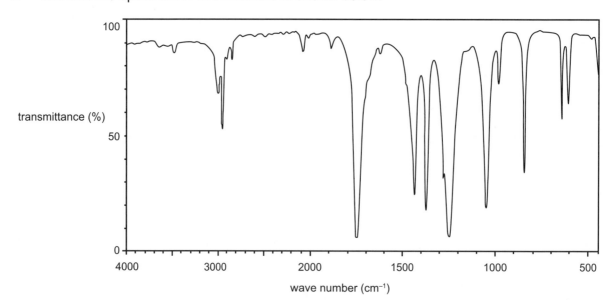

Data: SDBSWeb; http://sdbs.db.aist.go.jp, National Institute of Advanced Industrial Science and Technology

© Insight Publications

i. Is this molecule likely to be an alcohol? Justify your answer.   1 mark

_____

ii. Does this molecule contain a C=O bond? Justify your answer.   1 mark

_____

The proton NMR spectrum of the molecule is shown below. The peaks on this spectrum are:

- shift 9.8 triplet
- shift 2.4 quintet
- shift 1.1 triplet.

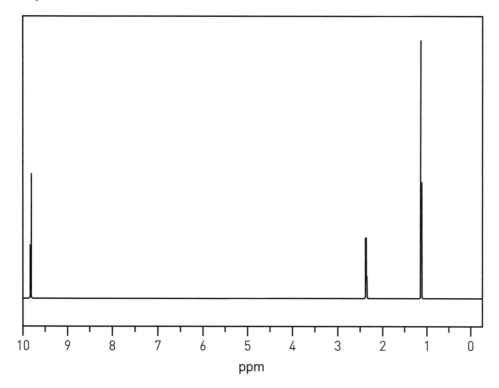

ppm

Data: SDBSWeb; http://sdbs.db.aist.go.jp, National Institute of Advanced Industrial Science and Technology

d. Use the proton NMR spectrum to identify the liquid. Justify your choice.   2 marks

_____

_____

Question 9 (11 marks)

A chemist has a sample of a liquid that is known to be either propan-1-ol or ethanoic acid. The liquid is tested with a mass spectrometer and the spectrum shown below is obtained.

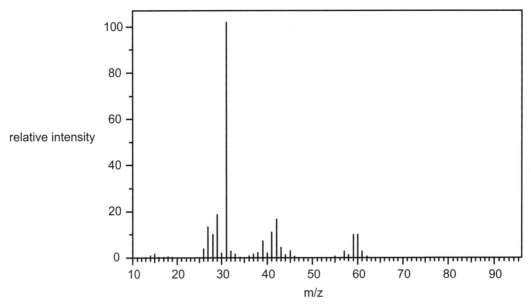

Data: SDBSWeb; http://sdbs.db.aist.go.jp, National Institute of Advanced Industrial Science and Technology

a. Identify whether propan-1-ol or ethanoic acid produced this spectrum. Justify your answer.   3 marks

_____

_____

_____

_____

A proton NMR is conducted on a liquid with molecular formula $C_3H_6O_2$ to determine its identity.

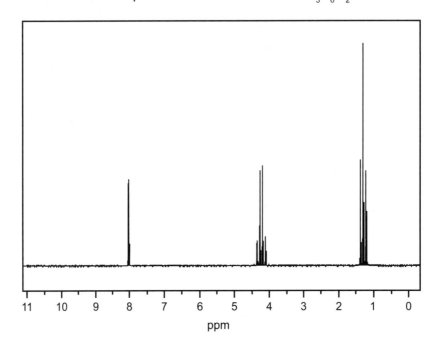

Data: SDBSWeb; http://sdbs.db.aist.go.jp, National Institute of Advanced Industrial Science and Technology

b. Use the spectrum provided above to name the molecule. Justify your answer.   3 marks

_____

_____

_____

The infrared spectrum below is from a molecule that is known to be one of either propanoic acid, methyl ethanoate or propan-1-ol.

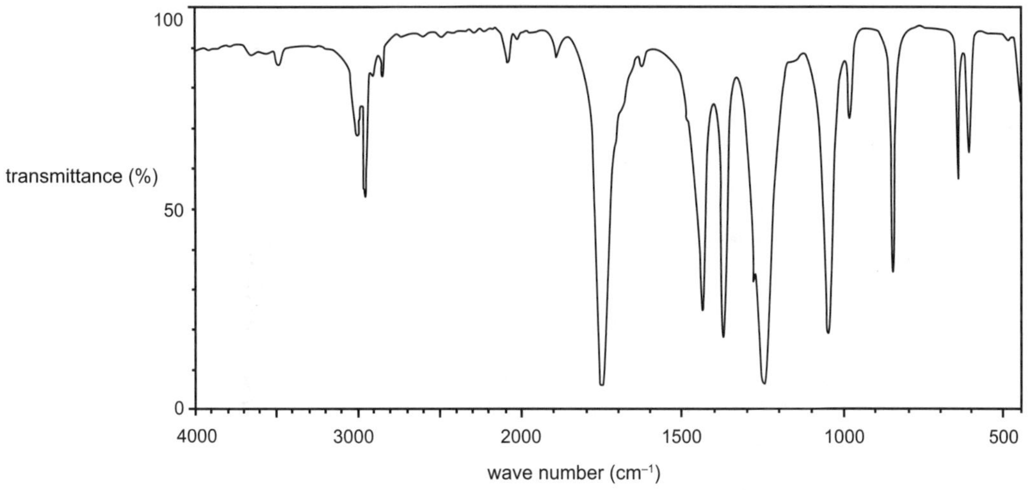

transmittance (%)

wave number (cm^{-1})

Data: SDBSWeb; http://sdbs.db.aist.go.jp, National Institute of Advanced Industrial Science and Technology

c. Use the spectrum above to identify the molecule. Justify your answer.   3 marks

_____

_____

_____

_____

d. A contaminated sample of water is known to contain propan-1-ol, hexan-2-ol and pentan-1-amine. The sample is injected into a HPLC column that uses ethanol as a solvent and a non-polar stationary phase.

Which of the three molecules will have the shortest retention time? Justify your answer.   2 marks

_____

_____

_____

Question 10 (12 marks)

A chemist isolates a molecule found as a contaminant in a sample of vinegar. To identify the molecule, the chemist tests the molecule with a number of instruments. The empirical formula is determined to be C_2H_4O.

a. The mass spectrum of the molecule is shown below.

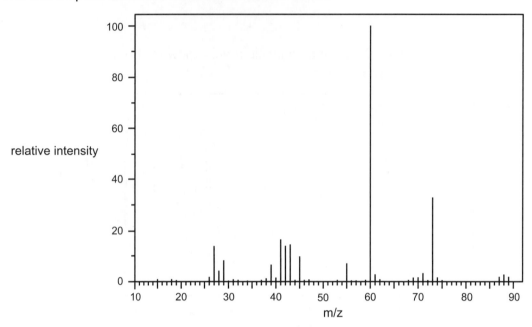

Data: SDBS Web, <http://sdbs.db.aist.go.jp>, National Institute of Advanced Industrial Science and Technology

i. What is the molecular formula of the molecule? Explain why.   2 marks

_____

_____

ii. Write an equation for the formation of the parent molecular ion in the ionisation area of the mass spectrum.   1 mark

_____

iii. Draw three possible structures for the molecule.   3 marks

b.   The infrared spectrum of the molecule is shown below.

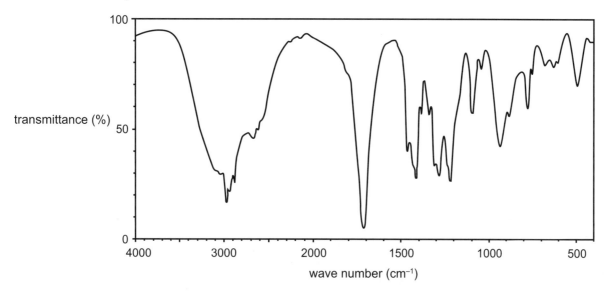

Data: SDBS Web, <http://sdbs.db.aist.go.jp>, National Institute of Advanced Industrial Science and Technology

Using the spectrum above, identify the contaminant molecule.
Justify your answer.   2 marks

_____

_____

_____

c.   The 1H NMR of the molecule is shown below.

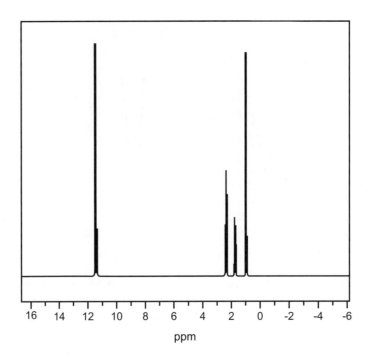

Data: SDBS Web, <http://sdbs.db.aist.go.jp>, National Institute of Advanced Industrial Science and Technology

The splitting pattern on this spectrum is difficult to read.

Referring to the structure of the molecule, explain which part caused each
set of peaks and what the splitting pattern might be.   4 marks

_____

_____

_____

_____

_____

_____

_____

Question 11 (6 marks)

Gluthathione is a tripeptide with an important role to play in the human body. It is formed in the liver and it acts as an antioxidant to prevent deterioration of many components of our body cells. The sequence of amino acids in this tripeptide is Glu-Cys-Gly.

a. Draw the structural formula this tripeptide.   3 marks

b. Specify the type of tertiary bonding that is likely to occur

    i. with glutamic acid.   1 mark

    ii. with cysteine.   1 mark

    iii. with glycine.   1 mark

Question 12 (12 marks)

a. A sample of threonine is dissolved in a dilute solution of NaOH.

    i. Draw the structure of threonine as it will exist in an alkaline solution.   1 mark

    ii. How many chiral carbon atoms does threonine have?   1 mark

b. Methanol is highly toxic in the human body due to its reaction with the enzyme alcohol dehydrogenase in the liver. Ethanol can be used to treat this problem owing to its action as a competitive enzyme inhibitor.

    i. Suggest a reason why ethanol might work as a competitive inhibitor to methanol.   1 mark

    ii. Explain the mechanism by which ethanol will work in this example.   2 marks

c. L-ascorbic acid can be used to treat scurvy. D-ascorbic acid has no impact on scurvy.

    i. How does L-ascorbic acid differ from D-ascorbic acid?   1 mark

    ii. Explain why the properties of the two molecules might cause them to act differently in the body.   2 marks

d. The two molecules shown below each have medicinal properties and can be extracted from plants.

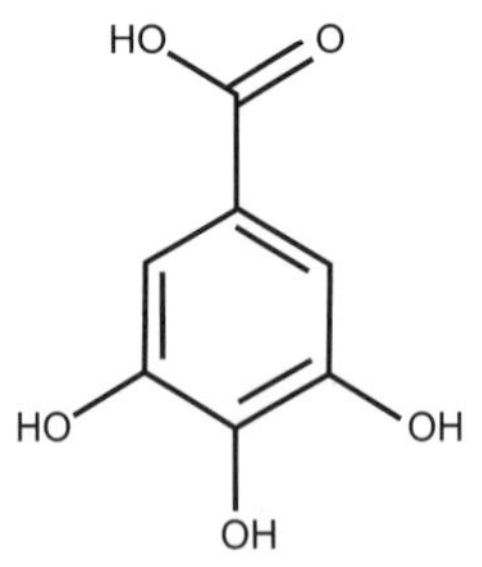

gallic acid 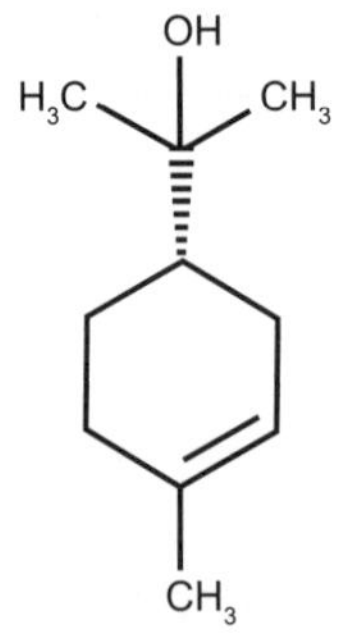   terpineol

    i. Describe how a gallic acid sample can be extracted from a plant.   1 mark

ii. The solvent used to extract gallic acid differs from the solvent used to extract terpineol. Explain how a suitable solvent is chosen for a particular sample.   2 marks

_____

_____

iii. A form of chromatography is usually used after a sample is extracted from a plant. Explain why.   1 mark

_____

_____

Unit 4 | Area of Study 3 How is scientific inquiry used to investigate the sustainable production of energy and/or materials?

Multiple choice

Question 1

A student wishes to investigate the rate of reaction between calcium carbonate and hydrochloric acid. She prepares five HCl solutions of concentrations, ranging from 0.1 M to 0.5 M. She adds 50 mL of each solution to a different beaker and places the five beakers on a hot plate set to 50 °C. Once the contents of the beaker have reached 50 °C, she adds 1.0 g of calcium carbonate to each beaker and records the time taken for the reaction to cease.

Select the option that correctly identifies the variables in this experiment.

	Control variable	Independent variable	Dependent variable
A.	temperature	time for reaction	HCl concentration
B.	temperature	HCl concentration	time for reaction
C.	HCl concentration	mass of $CaCO_3$	time for reaction
D.	time for reaction	HCl concentration	mass of $CaCO_3$

Question 2

Which one of the following statements about errors is the most correct?

A. Repeated trials can serve to minimise the impact of random errors.

B. Selecting an inappropriate indicator for a titration is an example of a random error.

C. A balance that has not been calibrated will cause random errors.

D. Repeated trials can serve to minimise the impact of systematic errors.

Question 3

The following graph represents the results from an experiment investigating the action of an enzyme, where temperature is the independent variable.

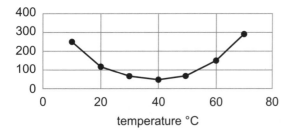

The dependent variable in this experiment could be the

A. number of collisions occurring per minute.

B. time taken for a positive result.

C. rate of the reaction.

D. number of successful collisions occurring per minute.

Unit 4, Area of Study 3: How is scientific inquiry used to investigate the sustainable production of energy and/or materials?

85

Short answer

Question 1 (6 marks)

Chemistry students frequently refer to the electrochemical series when answering questions relating to redox reactions.

Two chemistry students conduct experiments with half-cells to attempt to replicate the voltages of three of the common half-equations shown in elechtrochemical series.

The three half-cells used by the students are shown below. Over a number of days, the students test each combination possible in a galvanic cell and record the polarities of the electrodes and the voltages obtained. The students then use the voltages obtained to form their own mini-electrochemical series. For their series, the students make the nickel half-cell their standard and assign it a value of 0.00 V.

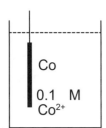
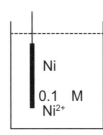
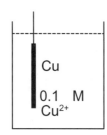

Cell combination	Voltage obtained	Positive electrode
cobalt–nickel	0.10 V	cobalt
cobalt–copper	0.41 V	copper
nickel–copper	0.51 V	copper

The students' electrochemical series was as follows.

Half-equation	V
$Cu^{2+}(aq) + 2e^- \rightleftharpoons Cu$	0.51
$Co^{2+}(aq) + 2e^- \rightleftharpoons Co$	0.10
$Ni^{2+}(aq) + 2e^- \rightleftharpoons Ni$	0.00

Their conclusion:

It is not possible in a school laboratory to obtain the same sequence as a Data Book electrochemical series.

a. The students' description of their experimental procedure contains some errors, which may include omissions.

   Briefly describe two errors (or omissions) in the students' procedure and explain how the procedure could be modified to improve the data obtained.   4 marks

   Error 1: _____

   _____

   _____

Error 2: _____

_____

_____

b.   The students arbitrarily assign the nickel half-cell a voltage of 0.00 Volts. Use the electrochemical series provided in your Data Book to determine the voltages that the other two half-cells should have if nickel has a value of 0.00 V.   2 marks

$Cu^{2+}(aq) + 2e^- \rightleftharpoons Cu$: _____

$Co^{2+}(aq) + 2e^- \rightleftharpoons Co$: _____

Question 2 (11 marks)

A student conducts an investigation of the rate of the reaction between magnesium and hydrochloric acid (HCl).

The details of the experiment are outlined below.

Hypothesis

The rate of reaction will be directly proportional to the concentration of the hydrochloric acid.

Design

The student prepares a solution of 1.0 M HCl and uses pipettes and deionised water to make a range of dilutions to this solution.

Five 250 mL flasks are assembled, ready to be attached to five gas syringes.

Using a beaker, 50.0 mL of solution is added to each flask.

A 1 cm piece of magnesium is added to each flask and a stopper inserted quickly so that the gas generated is trapped in the gas syringe.

The time taken for the volume collected to reach 20 mL is recorded.

Results

Concentration of HCl (M)	Time (s)
0.2	140
0.4	60
0.6	35
0.8	15
1.0	7

Conclusion

The rate of reaction drops as the concentration increases.

Unit 4, Area of Study 3: How is scientific inquiry used to investigate the sustainable production of energy and/or materials?

87

a. Write a balanced equation for the reaction occurring. 1 mark

b. **i.** Identify the independent variable. 1 mark

ii. Identify the dependent variable. 1 mark

iii. Identify two controlled variables. 2 marks

c. Identify two steps in the procedure that limit the accuracy of the results. 2 marks

d. Is the student's conclusion valid? Justify your answer. 2 marks

e. The student's results are significantly influenced by the action of a variable that the student thought was controlled.

Identify that variable and explain how it has influenced the results obtained. 2 marks

Question 3 (8 marks)

A student sets up a galvanic cell to investigate the relationship between the current in a cell and the concentration of the solution. A representation of the cell is shown below. The student's notes are also shown.

Hypothesis

The current in a cell is proportional to the concentration of the solution.

Cell

$Zn^{2+}(aq)$, $Zn(s)$ half-cell connected to a $Cu^{2+}(aq)$, $Cu(s)$ half-cell.

Design

50 mL of 1.0 M solution is added to both beakers.

Water is to be added in 5 mL increments to the copper half-cell to change the concentration. The current will be recorded after each increment.

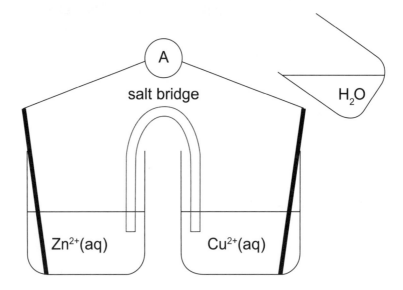

Volume of water added (mL)	Current (A)
0	0.78
5	0.82
10	0.86
15	0.89
20	0.91

Conclusion

The hypothesis is proven because the current increases as the water is added.

Unit 4, Area of Study 3: How is scientific inquiry used to investigate the sustainable production of energy and/or materials?

89

a. **i.** Identify the independent variable. 1 mark

_____

ii. Identify the dependent variable. 1 mark

_____

iii. Identify a controlled variable. 1 mark

_____

b. Is the student's conclusion consistent with the experimental data?
Justify your answer. 2 marks

_____

_____

_____

c. The student's teacher suggests that the experimental design needs to be modified in order to produce more valid data.

Identify a flaw with the experimental design and discuss how the student might address this issue. 3 marks

_____

_____

_____

_____

_____

_____

● Worked solutions

Unit 3 | Area of Study 1   What are the current and future options for supplying energy?

Multiple choice

Question 1

Answer: **C**

Explanatory notes

Option C is correct. Sugar cane is grown each year so it can be replenished within a reasonable time frame.

Option A is incorrect. Natural gas has finite reserves.

Option B is incorrect. Natural gas has finite reserves.

Option D is incorrect, as the killing of whales is politically unpopular and the number of whales would dwindle.

Question 2

Answer: **A**

Explanatory notes

Option A is correct. Australia is a dry continent, so the large-scale production of canola crops would place heavy demands on water, fertiliser and arable land. The farming equipment required will produce emissions and use significant amounts of energy.

Option B is incorrect. The oil for the biodiesel comes only from the seeds of the crop.

Option C is incorrect. The triglyceride in oil undergoes hydrolysis, then esterification or transesterification, rather than condensation.

Option D is incorrect. Each molecule of triglyceride produces three molecules of biodiesel.

Question 3

Answer: **D**

Explanatory notes

Option D is correct. If demand for electrical energy is unexpectedly high, the diesel generators can be switched on to meet the demand very quickly.

Option A is incorrect. Diesel generators produce CO_2 and a range of other pollutants.

Option B is incorrect because diesel is a petroleum product that is non-renewable.

Option C is incorrect. Many galvanic cells and fuel cells will be more efficient than a diesel generator.

Question 4

*Answer: **A***

Explanatory notes

Option A is correct. Alkanes have the general formula C_nH_{2n+2}. For the alkane molecule to have 38 atoms, the formula must be $C_{12}H_{26}$.

Option B is incorrect. It is an alkene.

Option C is incorrect because it contains too many atoms.

Option D is incorrect because it contains too many atoms.

Question 5

*Answer: **D***

Explanatory notes

Option D is correct. The activation energy is very low, and hence the reactants will be unstable. The enthalpy of the products is much lower than that of the reactants, so the reaction is highly exothermic.

Option A is incorrect. The low activation energy means the reactants are unstable and it is not endothermic.

Option B is incorrect. The reactants are not stable.

Option C is incorrect. The amount of energy released will be significant.

Question 6

*Answer: **B***

Explanatory notes

Option B is correct. The value of ΔH is $-940 - (-50) = -890$ kJ mol^{-1}. From the heats of combustion in the Data Book, this matches methane as a fuel (890 kJ mol^{-1}).

Option A is incorrect. Hydrogen has a different heat of combustion.

Option C is incorrect. Methanol has a different heat of combustion.

Option D is incorrect. Ethanol has a different heat of combustion.

Question 7

*Answer: **C***

Explanatory notes

Option C is correct. The reaction is exothermic but the value of ΔH is much lower than typical fuels, such as methane.

Option A is incorrect. The value of ΔH for Substance B is low and the high activation energy would also be a problem.

Option B is incorrect. The high activation energy of Substance B means it is unlikely to be unstable.

Option D is incorrect. The combustion of Substance B is exothermic.

Question 8

*Answer: **B***

Explanatory notes

Option B is correct.

$n(\text{methane}) = \dfrac{V}{V_m} = \dfrac{99.2}{24.8} = 4$ mol, which is the highest number of mole. Therefore, the energy released will be the highest.

Option A is incorrect. 50 g of methane is less than 4 mol.

Option C is incorrect. The number of mole of methane is less than option B.

Option D is incorrect. Ethane releases less energy per gram than methane.

TIP

» To best manage your time during a Chemistry exam, look for any chance to eliminate unlikely options in multiple-choice questions. A quick glance in the Data Book shows that ethane has a lower enthalpy per gram than methane, so option D can be eliminated quickly.

Question 9

*Answer: **B***

Explanatory notes

Option B is correct. Propan-1-ol will be relatively soluble due to the presence of an −OH group on a small molecule. The heat of combustion of propan-1-ol is consistent with the trend in heats of combustion of alcohol molecules listed in the Data Book.

Option A is incorrect. Pentane is insoluble in water. It also has a higher heat of combustion.

Option C is incorrect. Cyclohexane is insoluble in water.

Option D is incorrect. Stearic acid is insoluble due to its long non-polar hydrocarbon chain.

Question 10

*Answer: **B***

Explanatory notes

Option B is correct. 1.00 g of starch releases 16 kJ of energy.

The mass of triglyceride required to provide the same amount of energy is $\dfrac{1 \times 16}{37} = 0.432$ g.

Option A is incorrect.

Option C is incorrect. Triglyceride produces more energy per gram than carbohydrates.

Option D is incorrect.

Question 11

Answer: **C**

Explanatory notes

Option C is correct. The anaerobic digestor will produce biogas and the yeast tank will allow fermentation to produce bioethanol.

Option A is incorrect. Yeast will not produce biogas.

Option B is incorrect. The digestor will not produce bioethanol.

Option D is incorrect. Each tank has a different product.

Question 12

Answer: **A**

Explanatory notes

Option A is correct. The Data Book quotes the combustion figure for butane to be 2880 kJ mol^{-1}.

The heat of combustion per gram will be $= \dfrac{2880}{58} = 49.7$ kJ g^{-1}.

Option B is incorrect. This figure is the molar mass of butane.

Option C is incorrect. It gives the energy per mole, not per gram.

Option D is incorrect. The units are not kJ.

Question 13

Answer: **B**

Explanatory notes

Option B is correct.

$n(\text{butane}) = \dfrac{1.16}{58} = 0.020$ mol

$n(O_2) = \dfrac{13}{2} \times n(\text{butane}) = \dfrac{13 \times 0.02}{2} = 0.13$ mol

mass $O_2 = 0.13 \times 32 = 4.16$ g

Option A is incorrect. The figure does not take into account that oxygen gas exists as O_2.

Option C is incorrect. The figure is double the correct answer.

Option D is incorrect. It quotes the molar mass of oxygen atoms.

Question 14

Answer: **D**

Explanatory notes

Option D is correct.

$n(CO_2) = 4 \times n(C_4H_{10})$

$V = n \times 24.8 = 0.080 \times 24.8 = 1.98$ L

Option A is incorrect. Pressure has been converted to Pa.

Option B is incorrect. The correct answer is 1.98.

Option C is incorrect. The number of moles of CO_2 is $4 \times n(\text{butane})$.

Question 15

*Answer: **B***

Explanatory notes

Option B is correct. Energy produced by butane = 150 × 49.7 = 7460 kJ

$$n(\text{ethane}) = \frac{7460}{1560} = 4.78 \text{ mol}$$

$V = n \times 24.8 = 4.78 \times 24.8 = 119 \text{ L}$

Option A is incorrect. 24.8 L represents only 1 mole of ethane.

Option C is incorrect. 1560 represents only 1 mole of ethane.

Option D is incorrect. 7460 is the total amount of energy, not the volume of ethane.

Question 16

*Answer: **B***

Explanatory notes

Option B is correct. Energy released by cashew = $m \times \Delta H$ = 1.80 × 5.4 = 9.72 kJ = 9720 J

$$\Delta T \text{ for the calorimeter} = \frac{9720}{586} = 16.6 \text{ °C}$$

Option A is incorrect. It is half of the correct value.

Option C is incorrect. The correct answer is 16.6.

Option D is incorrect. It is twice the correct answer.

Question 17

*Answer: **B***

Explanatory notes

Option B is correct. If 100 g produces 1590 kJ, then 1 g produces 15.9 kJ = 15900 J.

15 900 = 4.18 × 500 × ΔT

$$\Delta T = \frac{15900}{4.18 \times 500} = 7.6 \text{ °C}$$

Option A is incorrect. The calculation should start with 1590 kJ rather than 555 kJ.

Option C is incorrect. It is double the correct answer.

Option D is incorrect. This calculation incorrectly uses 100 g of water rather than 500 g.

Question 18

*Answer: **C***

Explanatory notes

Option C is correct.

$$\text{Calibration factor} = \frac{VIt}{\Delta T} = \frac{5.4 \times 3.4 \times 5 \times 60}{64} = 861 \text{ J °C}^{-1}$$

Option A is incorrect. Time needs to be in seconds, not minutes.

Option B is incorrect. The answer is out by a factor of 10.

Option D is incorrect. It is double the correct answer.

Question 19

Answer: B

Explanatory notes

Option B is correct.

Energy = calibration factor × ΔT = 861 × 10.8 = 9300 J

Heat combustion = $\dfrac{9300}{1.12}$ = 8300 J g^{-1} = 8.30 kJ g^{-1}

Option A is incorrect. Answer is 8.30 kJ g^{-1}.

Option C is incorrect. The temperature change has not been used.

Option D is incorrect as well as the units

Question 20

Answer: B

Explanatory notes

Option B is correct. The oxidation states of sulfur are 2^- in H_2S, zero in S_8, 4^+ in SO_2 and $6+$ in SO_3.

Option A is incorrect. S_8 has an oxidation state of zero because it is an element.

Options C is incorrect. It lists in order of highest to lowest oxidation state.

Option D is incorrect. It has no sequence.

Question 21

Answer: D

Explanatory notes

Option D is correct. The oxidation number of the manganese atoms is reduced from 4+ to 2+. For reduction, the electrons must be accepted, and this is the case.

Option A is incorrect. It shows oxidation.

Option B is incorrect. Hydrogen gas is not produced in the overall equation.

Option C is incorrect. It is a repeat of the overall equation.

Question 22

Answer: C

Explanatory notes

Option C is correct. The correct formula for ethanol is C_2H_6O. The reaction is oxidation, so the electrons need to be on the right-hand side of the equation. The equation is balanced. The stem of the question contains a useful clue when it says 'in acidic conditions'. This phrase alerts you to look for the presence of H^+ ions.

Option A is incorrect. The formula for ethanol is wrong.

Option B is incorrect. It shows a reduction equation.

Option D is incorrect. The reaction does not occur in alkaline conditions and hydrogen gas will not be produced.

Question 23

*Answer: **A***

Explanatory notes

Option A is correct. The spontaneous reaction will be between iodine and aluminium metal. The iodine will be reduced and the aluminium oxidised, therefore the metal is the reductant.

Option B is incorrect. Aluminium ions are not the oxidising agent.

Option C is incorrect. Iodine is reduced.

Option D is incorrect. Iodide ions are not oxidised.

Question 24

*Answer: **C***

Explanatory notes

Option C is correct. The phrase 'continuous supply of fuel' suggests a fuel cell. Of the fuel cells listed, methane is the most likely to be obtained in a sustainable fashion. It is the main component of biogas.

Option A is incorrect. This lists the reactants for a secondary cell, not a fuel cell.

Option B is incorrect. Aluminium is not considered a sustainable fuel, nor is it easy to produce as a continuous supply.

Option D is incorrect. Butane is difficult to obtain in a sustainable fashion.

Question 25

*Answer: **A***

Explanatory notes

Option A is correct. All atoms are balanced, and the number of electrons correctly balances the charges present.

Option B is incorrect. The number of electrons should be 2.

Option C is incorrect. Several atoms are not balanced.

Option D is incorrect. The electrons should be on the left side of the equation.

Question 26

*Answer: **B***

Explanatory notes

Option B is correct. There are several species such as water or Fe^{2+} that appear in more than one half-equation.

Option A is incorrect. The electrochemical series gives no indication of reaction rate.

Option C is incorrect. The exact voltage will depend upon concentration and temperature.

Option D is incorrect. The series in the Data Book is only a small cross-section of possible redox reactions.

Question 27

Answer: **C**

Explanatory notes

Option C is correct. The reaction has to be oxidation to occur at the anode. Option C is correctly balanced. Note also that the half-equation in option C produces H^+ ions. This is consistent with the acidic conditions mentioned in the question.

Option A is incorrect. It is the reaction at the cathode.

Option B is incorrect. O^{2-} ions are not present.

Option D is incorrect. This is the overall equation for this cell.

TIP

» Be aware that one or both half-equations for the cell might be found on the electrochemical series in your Data Book.

Question 28

Answer: **A**

Explanatory notes

Option A is correct. The reaction at the cathode needs to be a reduction reaction. It is the oxygen gas that is reduced and the formation of OH^- is consistent with an alkaline environment.

Option B is incorrect. It shows oxygen reacting in an acidic environment.

Option C is incorrect. It shows an oxidation reaction.

Option D is incorrect. It shows an oxidation reaction and would require an acidic environment.

Short answer

Question 1a.i.

Worked solution

Renewable fuel: can be replenished at a sustainable rate.

Mark allocation: 1 mark

- 1 mark for sustainable replenishment.

Question 1a.ii.

Worked solution

Anaerobic: oxygen-free environment. Reaction is conducted in an airtight reactor.

Mark allocation: 1 mark

- 1 mark for oxygen-free environment.

Question 1a.iii.

Worked solution

The main component of biogas is methane.

Explanatory notes

Biogas produces a mixture of gases, varying according to the feedstock and the conditions used. The main component is usually methane – the composition of biogas being anything up to about 75%. CO_2, water and nitrogen are examples of other gases usually present.

Mark allocation: 1 mark

- 1 mark for methane.

Question 1a.iv.

Worked solution

The methane is usually combusted; the combustion will produce CO_2 and other emissions.

Explanatory notes

Biogas is usually burnt in a generator. The combustion of methane will produce CO_2, H_2O and small amounts of other emissions.

Mark allocation: 1 mark

- 1 mark for stating that the combustion of methane produces CO_2 and other emissions.

Question 1b.i.

Worked solution

1 mole of glucose and 6 moles of oxygen has more chemical potential energy than 6 moles of CO_2 and 6 moles of H_2O.

Explanatory notes

Plants produce glucose during photosynthesis. This is an endothermic reaction that absorbs energy from the Sun. Therefore, glucose has more potential energy than the molecules it was made from.

Mark allocation: 1 mark

• 1 mark for glucose and oxygen having more potential energy.

Question 1b.ii.

Worked solution

plant cells

Explanatory notes

Plant cells use photosynthesis to convert CO_2 and water to glucose and oxygen gas.

Mark allocation: 1 mark

• 1 mark for plant cells.

Question 1b.iii.

Worked solution

muscle cells in animals

Explanatory notes

Cellular respiration is the process of oxidising glucose to CO_2 and water in muscle cells.

Mark allocation: 1 mark

• 1 mark for muscle cells.

Question 2a.i.

Worked solution

$C_6H_{12}O_6(aq) \rightarrow 2CH_3CH_2OH(aq) + 2CO_2(g)$

Mark allocation: 1 mark

• 1 mark for a correctly balanced equation with states.

Question 2a.ii.

Worked solution

Ethanol produced from sources such as sugar cane waste is considered renewable because the sugar crops are grown each year, providing the same raw material.

Explanatory notes

Fermentation takes glucose and converts it to ethanol and carbon dioxide. Fermentation occurs in anaerobic conditions (i.e. in the absence of oxygen) and is catalysed by yeast. The source of glucose can vary with the region; for example, potato peel is used in Ballarat at a small-scale plant for ethanol. Although CO_2 is consumed as plants grow, the use of bioethanol is not carbon neutral. The farm machinery used, the transport type used and other stages of processing all require the use of energy.

Mark allocation: 1 mark

- 1 mark for explaining that the source can be replenished in a reasonable time frame.

Question 2b.i.

Worked solution

$$C_2H_6O(l) + 3O_2(g) \rightarrow 2CO_2(g) + 3H_2O(l)$$

Mark allocation: 1 mark

- 1 mark for a balanced equation and states.

Note: C_2H_5OH is also acceptable.

Question 2b.ii.

Worked solution

$$n(\text{ethanol}) = \frac{10\,000}{46} = 217.4 \text{ mol}$$
$$E = n \times \text{heat of combustion} = 217.4 \times 1360 = 2.96 \times 10^5 \text{ kJ}$$

Explanatory notes

Complete combustion produces CO_2 and H_2O. Balance the carbon atoms first, then the hydrogen atoms and, finally, the oxygen atoms.

10 kg of ethanol is 10 000 g. Calculate the number of moles of ethanol. The Data Book provides a figure of 1360 kJ for each mole of ethanol combusted.

Mark allocation: 2 marks

- 1 mark for calculating the number of moles of ethanol.
- 1 mark for calculating the amount of energy.

Question 2c.i.

Worked solution

$$C_2H_6O(l) + 2O_2(g) \rightarrow 2CO(g) + 3H_2O(l)$$

Mark allocation: 1 mark

- 1 mark for a correctly balanced equation and states.

Question 2c.ii.

Worked solution

$T = 400 + 273 = 673$ K

$n(CO) = 2 \times n(ethanol)$

$n(ethanol) = 217.4$ (as per **Question 2b.ii.**)

$V = n \times V_m = 217.4 \times 24.8 = 5400$ L

Explanatory notes

Incomplete combustion can produce a range of products. Students must read the question carefully to see if it is explicit as to whether carbon or carbon monoxide is the main product.

The molar volume equation can then be used to determine the volume.

Mark allocation: 2 marks

- 1 mark for the correct conversion of temperature and pressure.
- 1 mark for the correct calculation of volume.

Note: Consequential marks will be awarded if an incorrect value for $n(ethanol)$ is used, as long as the method is correct.

Question 3a.i.

Worked solution

$CH_4(g) + 2O_2(g) \rightarrow CO_2(g) + 2H_2O(l)$

Explanatory notes

The complete combustion of methane will produce carbon dioxide gas and water.

Mark allocation: 1 mark

- 1 mark for a balanced equation, including states.

Note: The value of ΔH is not required, as the question does not ask for a thermochemical equation.

Question 3a.ii.

Worked solution

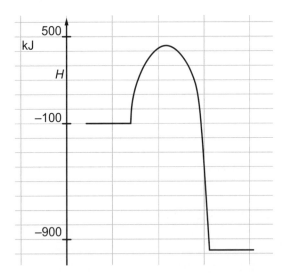

Explanatory notes

The combustion of methane will produce CO_2 and H_2O. The state of H_2O needs to be listed as liquid to match the Data Book, if conditions are at SLC and/or the quoted energy values are used. Always ensure that you take the context and reaction conditions into consideration when determining if H_2O is liquid or gas.

The value of ΔH for the combustion of methane is listed as -890 kJ mol^{-1}, making the enthalpy of the products -990 kJ. The activation energy of the exothermic reaction will be $1440 - 890 = 550$ kJ.

Mark allocation: 2 marks

- 1 mark for the correct shape of graph.

- 1 mark for the activation energy peaking at 450 kJ and enthalpy finishing at around -990 kJ.

Question 3b.i.

Worked solution

m (CH$_4$) = d × V = 0.66 g mL-1 × 50000 mL = 33000 g

Energy = 33000 × 55.6 (from Data Book) = 1.8×10^6 kJ or 1.8×10^3 MJ

Explanatory notes

The heat of combustion of methane is provided in the Data Book, in grams. Therefore, the first step is to calculate the mass of methane present. Note that it is easier to use the combustion value in kJ g^{-1} than the other value provided in the Data Book, kJ mol^{-1}.

Mark allocation: 1 mark

- 1 mark for the correct answer.

Question 3b.ii.

Worked solution

$n(\text{methane}) = \dfrac{33\,000}{16} = 2063 \text{ mol}$

$n(CO_2) = n(\text{methane}) = 2063 \text{ mol}$

$V = n \times V_m = 2063 \times 24.8 = 5.12 \times 10^4 \text{ L}$

Explanatory notes

The balanced equation shows that the number of mole of CO_2 will equal that of methane. The volume can be calculated using the molar volume gas equation.

Mark allocation: 3 marks

- 1 mark for the calculation of $n(CO_2)$.

- 2 marks for the correct substitution and calculation of volume, in L.

Question 4a.i.

Worked solution

$E = 4.18 \times 900 \times (24.4 - 18.8) = 21\,100 \text{ J} = 21.1 \text{ kJ}$

Explanatory notes

The energy required to heat the water is given by the formula $E = 4.18 \times m \times \Delta T$.

The units will be joules.

Mark allocation: 2 marks

- 1 mark for using the correct formula and temperature change.

- 1 mark for the correct answer and units.

Question 4a.ii.

Worked solution

$\Delta H = \dfrac{E}{m} = \dfrac{21\,100}{3.782 - 1.122} = 7920 \text{ J g}^{-1} = 7.92 \text{ kJ g}^{-1}$

Explanatory notes

To obtain the heat of combustion per gram, the energy obtained must be divided by the mass of biscuit burnt.

Mark allocation: 1 mark

- 1 mark for the correct answer.

Question 4b.

Worked solution

Possible modifications:

- put a lid on the beaker
- wrap the beaker in insulation
- change from using a beaker to using a bomb calorimeter
- use a metal can.

Explanatory notes

The heat losses from using an open glass beaker will be high. The design needs to be changed to limit these losses. A lid and insulation are obvious changes.

Mark allocation: 3 marks

- 1 mark for each valid design change listed (up to 3 marks).

Question 5a.

Worked solution

Methanol is the alcohol usually used to make biodiesel. The choice of ethanol to make Biodiesel A offers a more sustainable pathway. Ethanol can be produced on a large scale from petroleum or from waste organic matter. If the ethanol is sourced from byproducts of the wheat industry, for example, the production of this form of biodiesel could be sustainable and considered fully renewable. The detrimental impact on the environment would be less than the damaging impact from using the non-renewable alternative, petrodiesel.

Explanatory notes

Biodiesel is made from the reaction of a triglyceride with an alcohol. If the triglyceride and the alcohol are both sourced from organic waste matter, then the production will be a sustainable one. Methanol is often the alcohol used to produce biodiesel, and it is often sourced from petroleum. The shift to ethanol would be a positive move for the environment because all raw materials could be produced sustainably.

Mark allocation: 2 marks

- 1 mark for referring to the production of bioethanol.
- 1 mark for explaining how biodiesel is better for the environment than petrodiesel.

Question 5b.

Worked solution

$C_{10}H_{20}O_2(l) + 14O_2(g) \rightarrow 10CO_2(g) + 10H_2O(l)$

Explanatory notes

Complete combustion will form CO_2 and H_2O. The biodiesel must be a liquid because it does not form aqueous solutions.

Mark allocation: 2 marks

- 1 mark for the correct products.
- 1 mark for the correct coefficients and states.

Question 5c.

Worked solution

m(biodiesel) = $10 \times 12 + 20 \times 1 + 16 \times 2 = 172$ g mol^{-1}

n(biodiesel) = $\dfrac{1000}{172}$ = 5.81 mol

$n(CO_2)$ = $10 \times n$(biodiesel) = 58.1 mol

$m(CO_2)$ = $n \times M$ = 58.1×44 = 2560 g

Explanatory notes

This calculation involves several sequential steps.

1. Find the number of mole of biodiesel.
2. Use the mole ratio of biodiesel to CO_2 to find the number of mole of CO_2.
3. Use $m = n \times M$ to determine the volume of the CO_2 produced.

Mark allocation: 4 marks

- 1 mark for the biodiesel molar mass.
- 1 mark for the number of mole of biodiesel.
- 1 mark for the number of mole of CO_2.
- 1 mark for the correct use of ideal gas equation.

 TIP

» It is expected that you can complete numerical calculations correctly. The assessors do not have many topics they can use to assess these numerical skills. Therefore, expect that either the mass or volume of exhaust gas will need to be calculated or the concentration of a redox chemical will need to be determined from titration data.

Question 6a.i.

Worked solution

Biogas is gas generated from the decomposition of organic matter in an oxygen-free environment.

Mark allocation: 1 mark

- 1 mark for the correct definition.

Question 6a.ii.

Worked solution

Biogas is considered a renewable fuel because it can be replenished in a relatively short time.

Mark allocation: 1 mark

- 1 mark for the correct definition.

TIP

» It sounds very easy to explain what is a renewable fuel. However, students always struggle to find the right words. The fact that it can be replenished at a sustainable rate is usually a safe response. Stating that 'it can be recycled' is not a valid response.

Question 6b.i.

Worked solution

Reaction at the anode: $O_2(g) + 2H_2O(l) + 4e^- \rightarrow 4OH^-(aq)$

Explanatory notes

The reaction in a methane fuel cell will be between methane gas and oxygen gas. In an alkaline environment, the half-equations will use OH^- ions.

Mark allocation: 1 mark

- 1 mark for the correct half-equation.

Question 6b.ii.

Worked solution

Reaction at the cathode: $CH_4(g) + 8OH^-(aq) \rightarrow CO_2(g) + 6H_2O(l) + 8e^-$

Mark allocation: 1 mark

- 1 mark for the correct half-equation.

Question 6b.iii.

Worked solution

$CH_4(g) + 2O_2(g) \rightarrow CO_2(g) + 2H_2O(l)$

Explanatory notes

The overall equation is the same as that for the combustion of methane in air.

Mark allocation: 1 mark

- 1 mark for the correct equation.

Question 7a.i.

Worked solution

The oxidation state of silver is reduced from Ag^+ to zero.

Explanatory notes

The oxidation state of silver can be read from the electrochemical series.

Mark allocation: 1 mark

- 1 mark for the correct oxidation state change.

Question 7a.ii.

Worked solution

As silver is reduced, it will be the cathode.

Explanatory notes

Reduction is always at the cathode.

Mark allocation: 1 mark

- 1 mark for cathode.

Question 7a.iii.

Worked solution

silver

Explanatory notes

Reduction is at the cathode and the cathode is the positive terminal in a galvanic cell.

Mark allocation: 1 mark

- 1 mark for silver.

Question 7b.i.

Worked solution

Zinc metal is oxidised to zinc ions. You will see the electrode gradually crumble away.

Explanatory notes

The oxidation at the anode is $Zn(s) \rightarrow Zn^{2+}(aq)$. The zinc atoms are leaving the electrode, as they form zinc ions. You will see the size of the electrode slowly become smaller.

Mark allocation: 1 mark

- 1 mark for degrading the electrode.

Question 7b.ii.

Worked solution

Silver ions will come out of solution to form silver metal. The silver metal will increase the size of the electrode.

Explanatory notes

The reaction of Ag^+ ions to Ag metal will see the size of the electrode increase with more silver metal.

Mark allocation: 1 mark

- 1 mark for silver deposition.

Question 8a.

Worked solution

	Half-equation	Polarity
anode:	$Li \rightarrow Li^+ + e$	-ve
cathode:	$Fe^{4+} + 4e \rightarrow Fe$	+ve

Explanatory notes

Lithium metal is oxidised to Li^+ at the anode.

In FeS_2, iron has an unusual oxidation state of 4+. It is reduced to 0.

Mark allocation: 4 marks

- 1 mark for each correct half-equation.
- 1 mark for each correct polarity.

Question 8b.

Worked solution

Lithium metal reacts vigorously with water, making the cell dangerous to use.

Explanatory notes

Cells that use lithium metal must use conductive polymer electrolytes to prevent the reaction of lithium with water.

Mark allocation: 1 mark

- 1 mark for mentioning lithium's reactivity with water.

Question 8c.

Worked solution

A voltage of 1.5 V matches the voltage of conventional zinc–carbon batteries. Therefore the battery is compatible and has the advantage of being lighter.

Mark allocation: 1 mark

- 1 mark for a valid explanation.

Question 8d.

Worked solution

The battery will go flat when one or both reactants is all used up. In this case, it would be when the silver ion concentration is zero or the zinc electrode has completely collapsed.

Mark allocation: 1 mark

- 1 mark for a valid explanation.

Unit 3 | Area of Study 2 — How can the rate and yield of chemical reactions be optimised?

Multiple choice

Question 1

*Answer: **D***

Explanatory notes

Option D is correct. Each S atom in S_8 gains two electrons to form S^{2-} ions.

Option A is incorrect. The sulfur exists as S_8 as a reactant, and the electrons are on the wrong side of the half-equation.

Option B is incorrect. 16 electrons are required for each S_8 molecule reacting.

Option C is incorrect. 16 electrons are required, and the electrons are on the wrong side of the half-equation.

TIP

» It is common for the exam to use a new type of galvanic cell and expect you to be able to apply principles of redox chemistry to an unfamiliar context.

Question 2

*Answer: **D***

Explanatory notes

Option D is correct. The half-equations in this cell are

$$MnO_2(s) + Li^+ + e^- \rightarrow LiMnO_2(s)$$

$$Li(s) \rightarrow Li^+ + e^-$$

The overall equation is obtained by adding these half-equations together:

$$Li(s) + MnO_2(s) \rightarrow LiMnO_2(s)$$

It is possible to deduce the overall equation from the diagram rather than working through the two half-equations.

Option A is incorrect. Li_2MnO_2 is not present in the cell.

Option B is incorrect. Lithium metal is a reactant, whereas oxygen gas is not.

Option C is incorrect. Lithium metal is a reactant, not a product.

Question 3

Answer: C

Explanatory notes

Option C is correct. When going from MnO_2 to $LiMnO_2$, the manganese is changing from Mn^{4+} to Mn^{3+}.

Option A is incorrect. It represents the change at the anode. It is an oxidation reaction that will occur at the anode.

Option B is incorrect. The manganese ends up as Mn^{3+}.

Option D is incorrect. The manganese ends up as Mn^{3+}.

Question 4

Answer: B

Explanatory notes

Option B is correct. When the cell is being recharged, the anode is the positive electrode, where oxidation occurs.

Option A is incorrect. It is the discharge reaction at the cathode, as it is reduction.

Option C is incorrect. It will occur at the cathode.

Option D is incorrect. It is the anode discharge reaction.

Question 5

Answer: A

Explanatory notes

Option A is correct. The pH rises as the H^+ ions are converted to water as the cell discharges.

Option B is incorrect. Mass is conserved in a reaction.

Option C is incorrect. The products are insoluble.

Option D is incorrect. Some of the lead changes from zero to 2+.

Question 6

Answer: C

Explanatory notes

Option C is correct. A feature of rechargeable cells is that the products stay in contact with the electrodes.

Options A, B and D are incorrect. These reasons are irrelevant to rechargeability.

Question 7

Answer: A

Explanatory notes

Option A is correct. The polarity of the electrodes does not change when discharge is swapped for recharge. Because zinc is oxidised during discharge, it is the negative anode.

Options B and D are incorrect. The anode and cathode swap during discharge and recharge.

Option C is incorrect. During recharge the electrons flow in the opposite direction.

Question 8

Answer: C

Explanatory notes

Option C is correct. The discharge reaction is the spontaneous reaction between zinc and bromine to form zinc ions and bromide ions. The recharge equation is the reverse of this.

Option A is incorrect. It shows the discharge reaction.

Options B and D are incorrect. Both have a mixture of reactants and products on both sides of the equation.

Question 9

Answer: C

Explanatory notes

Option C is correct. Lithium is converted to Li^+ ions and manganese is reduced from 4+ to 3+.

Option A is incorrect. The electrons flow the other way.

Option B is incorrect. Lithium is the anode, not the cathode.

Option D is incorrect. It is the lithium metal that reacts, not the lithium ions.

Question 10

Answer: B

Explanatory notes

Option B is correct. The reaction can be found by adding together the two half-equations provided.

Option A is incorrect. The reaction it is not balanced.

Options C and D are incorrect. They show the lithium ions reacting, whereas it is the lithium metal that reacts.

Question 11

Answer: **C**

Explanatory notes

Option C is correct. The extra catalyst in Experiment 2 will lead to a faster reaction but it will not change the volume of oxygen released. (The amount of H_2O_2 decomposing has not changed.)

Option A is incorrect. The amount of H_2O_2 decomposing in both experiments is the same.

Option B is incorrect. MnO_2 is not a reactant; it is a catalyst.

Option D is incorrect. The greater amount of catalyst in Experiment 2 will lead to a faster rate.

Question 12

Answer: **B**

Explanatory notes

Option B is correct. The temperature is adjusted to 50 °C for each experiment; this is the controlled variable. The HCl concentration is the independent variable. The time taken for the reaction to cease is the dependent variable because it depends upon the HCl concentration.

Option A is incorrect. The dependent and independent variables are the wrong way around.

Option C is incorrect. The mass of $CaCO_3$ is not a variable.

Option D is incorrect. The time for the reaction is not controlled.

Question 13

Answer: **C**

Explanatory notes

Option C is correct. The initial rate of Reaction A is higher, which is consistent with an increase in the concentration of the HCl, leading to a greater number of successful collisions.

Option A is incorrect. A lower temperature would cause the initial rate of Reaction A to be lower than that of Reaction B.

Option B is incorrect. An increase in the mass of $CaCO_3$ would lead to a greater final volume of gas.

Option D is incorrect. The shape of the beaker will not alter any factors that may affect the rate, such as temperature, concentration or surface area.

Question 14

Answer: **B**

Explanatory notes

Option B is correct. The greater mass of $CaCO_3$ is consistent with the higher volume of CO_2 produced. The initial rate is also lower due to the large chip being used. A larger chip will have a lower surface area.

Option A is incorrect. A catalyst would lead to an increase in the reaction rate.

Option C is incorrect. It does not explain the higher volume of CO_2 released.

Option D is incorrect. It does not explain the higher volume of CO_2 released.

Question 15

Answer: **D**

Explanatory notes

Option D is correct. The products of a reaction are always listed in the numerator of an equilibrium expression, and the reactants are listed in the denominator. The products will both have a coefficient of ½ in the balanced equation.

Option A is incorrect. The equation would need to be doubled for option A to be the correct answer.

Option B is incorrect. The equation is the wrong way around.

Option C is incorrect. The equation is both the wrong way around and double what it should be.

Question 16

Answer: **C**

Explanatory notes

Option C is correct.

$$K = \frac{[M] \times [M]^4}{[M] \times [M]^2} = M^2$$

Option A is incorrect. The answer must be M^2.

Option B is incorrect. The answer must be M^2.

Option D is incorrect. The answer must be M^2.

Question 17

Answer: **B**

Explanatory notes

Option B is correct. If the pressure is decreased, the system will oppose this change by moving to the side with more particles. There are more products than reactants so the forward reaction is favoured, improving the yield.

Option A is incorrect. An increase in temperature will cause the system to shift in the endothermic direction in order to reduce the added heat. Therefore, the back reaction is favoured, decreasing the yield.

Option C is incorrect. A decrease in volume will increase the pressure and concentration. The back reaction is favoured in order to decrease the concentration of particles.

Option D is incorrect. The addition of CO_2 will favour the back reaction, to remove CO_2.

Question 18

Answer: **D**

Explanatory notes

Option D is correct. An increase in pressure will favour the products because the ratio of reactants to products is $4 : 2$. This is an application of Le Chatelier's principle – that reactions partially oppose any change from equilibrium.

Option A is incorrect. An increase in temperature for an exothermic reaction will favour the reverse reaction.

Option B is incorrect. An increase in volume is the same as a decrease in pressure, which would favour the reverse reaction.

Option C is incorrect. The addition of a catalyst will not change the yield.

Question 19

Answer: **A**

Explanatory notes

Option A is correct. If the temperature is increased in an exothermic reaction, the reverse reaction is favoured. The amount of reactants will increase in stoichiometric amounts and the amount of product will decrease. The amounts given in option A are consistent with the balanced equation.

Option B is incorrect. The ratios in the equation have not been taken into account.

Option C is incorrect. The amounts of all chemicals cannot increase.

Option D is incorrect. The system does not go in the forward direction.

Question 20

Answer: **C**

Explanatory notes

Option C is correct. The expression for the units for $K = \dfrac{[M]^4 \times [M]^6}{[M]^4 \times [M]^5} = M$.

Option A is incorrect. The answer must be M.

Option B is incorrect. This response indicates that the equilibrium expression has been applied incorrectly, with the reactants above the products.

Option D is incorrect. The answer must be M.

Question 21

Answer: **A**

Explanatory notes

Option A is correct. When the volume is halved, the concentrations all double. The back reaction is favoured as the system shifts to reduce the number of particles, but the concentration of NO will not drop as far as its original value at t_1.

Option B is incorrect. The amount of NO is lowered due to the back reaction being favoured.

Option C is incorrect. The amount of NO does change.

Option D is incorrect. The only variable that will change the value of K is temperature.

TIP

» **When applying Le Chatelier's principle, be sure to check exactly what quantity is being asked for. It could be:**

- **the position of equilibrium**
- **the impact on the yield**
- **the impact on K**
- **the impact on the amount of a substance**
- **the impact on the concentration of a substance.**

Question 22

Answer: **D**

Explanatory notes

Option D is correct. The addition of water decreases the number of collisions because the particles are farther apart. The reverse reaction is favoured because the ratio of reactants to products is 2 : 1. The intensity of red decreases and the amount of SCN⁻ ions increases. The concentration of SCN⁻ is lower because the volume of the solution has increased.

Option A is incorrect. The temperature has not changed, so the value of K will not change.

Option B is incorrect. The addition of water lowers the concentration of the SCN⁻ ions.

Option C is incorrect. The addition of water lowers the concentration of the SCN⁻ ions.

Question 23

Answer: A

Explanatory notes

Option A is correct. The possible oxidants are Cu and water. Cu is the strongest reducing agent present so it will react to form Cu^{2+} ions in solution.

Option B is incorrect. It is a reduction reaction that might occur at a cathode.

Option C is incorrect. Cu reacts in preference to water.

Option D is incorrect. It is the reaction that will occur at the cathode.

Question 24

Answer: C

Explanatory notes

Option C is correct. Cu^{2+} ions will form at the anode at the same rate they are converted to Cu metal at the cathode.

Option A is incorrect. It ignores the reaction at the cathode.

Option B is incorrect. The concentration of SO_4^{2-} will not change.

Option D is incorrect. It ignores the anode reaction.

Question 25

Answer: D

Explanatory notes

Option D is correct. Zinc ions are the strongest oxidising agent present and react to form zinc metal at the cathode. Copper metal is the strongest reducing agent present. It reacts to form copper(II) ions at the anode.

Option A is incorrect. Oxygen gas will not form. Oxidation of water to oxygen is non-spontaneous and would require energy.

Option B is incorrect. Hydrogen and oxygen gases do not form. Water is not electrolysed because no energy is being supplied and zinc ions and copper metal are, respectively, the strongest oxidant and reductant present.

Option C is incorrect. The zinc ions react at the cathode.

Question 26

Answer: **C**

Explanatory notes

Option C is correct.

$Q = It = 10 \times 193 = 1930$ C

$n(e) = \dfrac{1930}{96500} = 0.02$ mol

The number of mole of metal obtained has to be consistent with 0.02 mol and a balanced half-equation.

To identify the metal, we need to know its molar mass. From options C and D remaining, we can see that $n(\text{metal}) = n(e)$.

$M(\text{metal}) = \dfrac{m(\text{metal})}{n(\text{metal})} = \dfrac{0.138 \text{ g}}{0.02 \text{ mol}} = 6.9$ g mol^{-1}

Therefore, the metal is lithium.

Option A is incorrect. It is a reaction that would occur at the anode, not the cathode.

Option B is incorrect. The equation is the wrong way around.

Option D is incorrect. The mass of sodium would not be consistent with the calculated molar mass.

Question 27

Answer: **A**

Explanatory notes

Option A is correct.

$n(\text{gas})$ at standard laboratory conditions (SLC) $= \dfrac{0.248}{24.8} = 0.01$ mol

$n(\text{gas})$ is therefore half the number of mole of electrons calculated in **Question 26**.

The gas must have a ratio of $1:2$ between the number of mole of gas and the number of mole of electrons. $2Cl^-(l) \rightarrow Cl_2(g) + 2e^-$ has the correct ratio.

Option B is incorrect. The ratio between oxygen and electrons would be $1:4$.

Option C is incorrect. The ratio between carbon dioxide and electrons would be $1:4$.

Option D is incorrect. The ratio between nitrogen and electrons would be $1:6$, as the equation would be $2N^{3-} \rightarrow N_2 + 6e^-$.

Question 28

Answer: **C**

Explanatory notes

Option C is correct.

$Q = It = 9650 \times 90 = 869\,000$ C

$n(\text{Ag obtained}) = \dfrac{869\,000}{96\,500} = 9$ mol

This is a higher number of mole than any other option.

Option A is incorrect. No metal is produced from the electrolysis of an aqueous aluminium ion solution.

Option B is incorrect. The time would need to be 180 seconds for the amount of copper to match the amount of silver produced.

Option D is incorrect. Although the amount of charge is the same as that of option B, the 3+ oxidation state of aluminium ions means that less metal is obtained.

 TIP

> » It is common for electrolysis questions to include aqueous solutions of reactive metals – for example, option A refers to $AlCl_3$(aq). These solutions will not produce any metal because water is reduced instead of the metal ion. Watch for reactive metals that have a voltage lower than water (–0.83 V) on the electrochemical series.

Question 29

Answer: **A**

Explanatory notes

Option A is correct. Reduction occurs at the cathode. In the electrolysis of molten $MgCl_2$, the reduction reaction is of magnesium ions to magnesium metal.

Option B is incorrect. It is not a balanced half-equation.

Option C is incorrect. It is an oxidation reaction.

Option D is incorrect. The electrolyte is not aqueous and the half-equation is not a reduction reaction.

Question 30

Answer: **B**

Explanatory notes

Option B is correct.

$Q = It = 12600 \times 60 \times 60 = 4.54 \times 10^7 \text{ C}$

$n(e) = \dfrac{4.54 \times 10^7}{96500} = 470 \text{ mol}$

$n(Cl_2) = n(e)/2 = 235 \text{ mol}$

volume of Cl_2 at SLC $= n \times 24.8 = 5800 = 5.8 \times 10^3 \text{ L}$

Option A is incorrect. It would be correct if the gas produced was oxygen.

Option C is incorrect. The number of mole of chlorine produced is half the number of mole of electrons.

Option D is incorrect. The answer is $5.8 \times 10^3 \text{ L}$.

Question 31

Answer: **C**

Explanatory notes

Option C is correct. The graph for sodium or lithium can be used to establish values.

One mole of sodium weighs 23 g. This information allows you to discern that one mole of electrons is producing around 40 g of the third metal. Each metal has to be tested to see if this data matches it. Potassium has a charge of +1, so 1 mole of electrons will produce 1 mole of potassium, which is 39 g. A value of 39 matches that of potassium.

Option A is incorrect. The molar mass of silver is far greater than 39.

Option B is incorrect. Magnesium has a charge of +2 so 1 mole of electrons would produce 0.5 mole of magnesium and the mass would be around 12.2 g, not 39.

Option D is incorrect. The charge on calcium is +2 and half a mole of calcium has a mass of 20 g not 39 g.

Question 32

Answer: **B**

Explanatory notes

Option B is correct. Lithium and sodium ions have the same oxidation state, so the number of mole of each will be the same.

$Li^+ + e^- \rightarrow Li$

$Na^+ + e^- \rightarrow Na$

Option A is incorrect. Metals with a different oxidation state will produce a different number of mole of metal.

Option C is incorrect. The gradient will be the same.

Option D is incorrect. The gradient will be the same.

Short answer

Question 1a.

Worked solution

i. Anode: $Zn(s) + 2OH^-(aq) \rightarrow ZnO(s) + H_2O(l) + 2e^-$

ii. Cathode: $O_2(g) + 2H_2O(l) + 4e^- \rightarrow 4OH^-(aq)$

Explanatory notes

For a spontaneous reaction, oxygen reacts with zinc metal. The second half-equation must be reversed to show it is zinc reacting. The reaction of oxygen is reduction, which will occur at the cathode. Oxidation occurs at the anode.

Mark allocation: 2 marks

- 1 mark for a correct half-equation with states for the anode.
- 1 mark for a correct half-equation with states for the cathode.

Note: It is also correct for the coefficients in the cathode half-equation to be halved.

Question 1b.i.

Worked solution

$2Zn(s) + O_2(g) \rightarrow 2ZnO(s)$

Explanatory notes

The overall equation is obtained by combining the two half-equations. The zinc half-equation is doubled to balance the number of electrons. It is convention to have all whole-number coefficients.

Mark allocation: 1 mark

- 1 mark for a correctly balanced equation with states included.

Question 1b.ii.

Worked solution

$0.34 - (-1.25) = 1.59$ V

Explanatory notes

The cell voltage is the difference between the potentials of the two half-equations. The lower potential is subtracted from the higher potential to calculate the difference.

Mark allocation: 1 mark

- 1 mark for the correct answer of 1.59 V.

Question 1c.i.

Worked solution

Zinc is relatively abundant.

Oxygen can be obtained from air.

Explanatory notes

The zinc–air cell is inexpensive because the reactants are cheap and abundant. Oxygen can be obtained from the air and zinc is mined in many different countries. Graphite electrodes are also inexpensive.

Mark allocation: 2 marks

- 1 mark for each valid reason supplied (up to 2 marks).

Question 1c.ii.

Worked solution

Fuel cells rely on a continuous supply of reactants. Regularly changing the zinc electrode provides a continuous supply of zinc and there is a continuous supply of air.

Explanatory notes

This cell is equivalent to a fuel cell in that it can run indefinitely if zinc is continually replaced. It is not recharged.

Mark allocation: 1 mark

- 1 mark for noting a continuous supply of reactants.

Question 2a.i.

Worked solution

Ni^{3+} to Ni^{2+}

Explanatory notes

Nickel goes from $2NiO(OH)$ to $Ni(OH)_2$. Knowing that the charge on oxygen ions is usually −2 and on hydroxide ions is usually −1, the oxidation state of nickel will change from +3 to +2.

Mark allocation: 1 mark

- 1 mark for Ni^{3+} to Ni^{2+}.

Note: The answer could be as simple as 3+ to 2+, although an answer such as 'decreases by 1+' does not give a sufficient reason to back up the answer.

Question 2a.ii.

Worked solution

Cadmium, because its oxidation number increases from zero to 2+.

Explanatory notes

Cadmium metal is replacing nickel ions in this cell, therefore cadmium is lower on the electrochemical series and is a stronger reducing agent.

Mark allocation: 1 mark

- 1 mark for cadmium.

Question 2a.iii.

Worked solution

$Cd(s) + 2OH^-(aq) \rightarrow Cd(OH)_2(s) + 2e^-$

Explanatory notes

The overall equation shows cadmium metal reacting to form $Cd(OH)_2$. This reaction of Cd to Cd^{2+} will involve the release of two electrons.

Mark allocation: 1 mark

- 1 mark for the correct half-equation (with the OH^- ions included).

Note: States can be ignored.

Question 2b.i.

Worked solution

Positive electrode: $Ni(OH)_2(s) + OH^-(aq) \rightarrow NiO(OH)(s) + H_2O(l) + e^-$

Negative electrode: $Cd(OH)_2(s) + 2e^- \rightarrow Cd(s) + 2OH^-(aq)$

Explanatory notes

When a cell is recharging, the half-equations are the reverse of the discharge equations. The polarity of the electrodes will not change between discharge and recharge.

Mark allocation: 2 marks

- 1 mark for each equation (up to 2 marks).

Note: States can be ignored.

Question 2b.ii.

Worked solution

a little over 1.2 V

Explanatory notes

The voltage used in a recharger needs to be slightly higher than the discharge voltage in order to push the reactions in the opposite direction. The voltage should not be too high however, or side reactions might occur, making unwanted products.

Mark allocation: 1 mark

- 1 mark for a voltage over 1.2.

Question 2b.iii.

Worked solution

Electrons travel from the anode to the cathode in both cells.

Explanatory notes

Fuel cells and rechargeable batteries are both galvanic cells that produce electric current from spontaneous chemical reactions. Electrons will flow from the negative anode to the positively charged cathode.

Mark allocation: 1 mark

- 1 mark for a valid answer.

Note: The answer could refer to the polarity, non-disposability or the direction of electron flow.

Question 2c.

Worked solution

Anode half-equation: $CH_4(g) + 2H_2O(l) \rightarrow CO_2(g) + 8H^+(aq) + 8e^-$

Cathode half-equation: $O_2(g) + 4H^+(aq) + 4e^- \rightarrow 2H_2O(l)$

Overall equation: $CH_4(g) + 2O_2(g) \rightarrow CO_2(g) + 2H_2O(l)$

Explanatory notes

The overall equation in a fuel cell is the same as that for combustion in air; that is, methane and oxygen reacting to form CO_2 and H_2O.

The oxygen half-equation in acidic conditions will be the one at +1.23 on the electrochemical series. The methane will form CO_2 and hydrogen ions.

Mark allocation: 3 marks

- 1 mark for each correct equation (up to 3 marks).

Note: States must be included.

Question 3a.i.

Worked solution

	Half-equation	Polarity
Anode	$Zn(s) \rightarrow Zn^{2+}(aq) + 2e^-$	–ve
Cathode	$Br_2(l) + 2e^- \rightarrow 2Br^-(aq)$	+ve

Explanatory notes

Both half-equations are provided in the electrochemical series in the Data Book. When the cell discharges, the reaction is between bromine liquid and zinc metal. The reaction of bromine is a reduction reaction, which will occur at the cathode. The cathode in a galvanic cell is positive.

Mark allocation: 3 marks

- 1 mark for each correct half-equation; states are not required (up to 2 marks).

- 1 mark for both correct polarities.

Question 3a.ii.

Worked solution

$Zn(s) + Br_2(l) \rightarrow 2Br^-(aq) + Zn^{2+}(aq)$

Explanatory notes

The overall reaction is between zinc metal and bromine liquid to form zinc ions and bromide ions.

Mark allocation: 1 mark

- 1 mark for a correctly balanced equation with states.

Question 3a.iii.

Worked solution

	Equation	Polarity
Anode	$2Br^-(aq) \rightarrow Br_2(l) + 2e^-$	+ve
Cathode	$Zn^{2+}(aq) + 2e^- \rightarrow Zn(s)$	−ve

Explanatory notes

To recharge the cell, electrolysis is used to re-form the zinc and bromine. The reaction of bromide ions is an oxidation reaction that occurs at the anode. The anode in an electrolytic cell is positive.

Mark allocation: 3 marks

- 1 mark for each correct half-equation; states are not required (up to 2 marks).
- 1 mark for both correct polarities.

Question 3b.i.

Worked solution

predicted voltage = 1.09 − (−0.76) = 1.85 V

Explanatory notes

The predicted voltage will be equal to the difference between the voltages of each half-equation given on the electrochemical series. The difference between the bromine and zinc half-equations is 1.85 V.

Mark allocation: 1 mark

- 1 mark for the correct voltage.

Question 3b.ii.

Worked solution

The values given on the electrochemical series assume standard conditions. This cell might not be operating at standard conditions.

Explanatory notes

The actual voltage obtained by a cell is often less than the predicted value because conditions are not standard and cells have internal resistance.

Mark allocation: 1 mark

- 1 mark for a valid reason, such as non-standard conditions or internal resistance in the cell.

Question 3c.

Worked solution

$n(Zn) = \dfrac{520}{65.4} = 7.95$ mol

$n(e) = 2n(Zn) = 2 \times 7.95 = 15.9$ mol

$Q = n(e) \times 96500 = 15.9 \times 96500 = 1.53 \times 10^6$ C

$t = \dfrac{Q}{I} = \dfrac{1.53 \times 10^6}{4.8} = 3.2 \times 10^5$ s

Explanatory notes

Zinc metal is oxidised in this cell. If all the zinc reacts, the number of mole reacting is 7.95 mol. Every atom of zinc oxidised releases two electrons, allowing the amount of charge to be calculated as well as the time it takes for this reaction to occur.

Mark allocation: 3 marks

- 1 mark for calculating the number of mole of electrons.
- 1 mark for calculating the charge.
- 1 mark for calculating the time; the answer must be to two significant figures.

 TIP

> » If your answer to this question has more than two significant figures, you are likely to lose the final mark. The value of 4.8 A given in the question has only two significant figures, so the answer should match. The number of significant figures in the answer is dictated by the item of data that has the fewest number of significant figures.

Question 4a.

Worked solution

$2Li(s) + 2H_2O(l) \rightarrow 2LiOH(aq) + H_2(g)$

Lithium reacts explosively with water. Production of hydrogen gas in an enclosed object is also a problem.

Explanatory notes

Lithium reacts explosively with water. This is a dangerous reaction and the formation of explosive hydrogen gas presents further issues.

Mark allocation: 2 marks

- 1 mark for the equation and 1 mark for referring to either the explosive reaction of lithium with water or the presence of hydrogen gas.

Question 4b.

Worked solution

Anode: $Li \rightarrow Li^+ + e$

Cathode: $S + 2e \rightarrow S^{2-}$

Explanatory notes

Lithium metal is reacting to form lithium ions. The half-equation for this reaction is listed on the electrochemical series in the Data Book. This reaction is oxidation, therefore it occurs at the anode.

Sulfur reacts at the cathode to form S^{2-} ions. The charge on sulfur can be deduced from the formula of Li_2S.

Mark allocation: 2 marks

- 1 mark for each correct equation.

Question 4c.

Worked solution

$8Li_2S \rightarrow 16Li + S_8$

Explanatory notes

The recharge equation is the reverse of the discharge equation.

Mark allocation: 1 mark

- 1 mark for reversing the equation supplied in the question.

Question 4d.

Worked solution

$Q = It = 2.5 \times 15 \times 60 = 2250$ coulomb

$n(e) = \dfrac{Q}{96500} = 0.0233$ mol

$n(Li) = n(e) = 0.0233$ mol

$m(Li) = n \times M = 0.0233 \times 6.9 = 0.16$ g

Explanatory notes

The charge can be calculated from the current and time.

The amount of charge provides the number of mole of electrons.

Lithium ions have a charge of +1 so the number of mole of lithium equals the number of mole of electrons.

Mark allocation: 3 marks

- 1 mark for calculating the charge.
- 1 mark for calculating the number of mole of lithium.
- 1 mark for calculating the mass.

Question 5a.

Worked solution

The volume of hydrogen gas is doubled so the mass of magnesium must have been doubled.

Explanatory note

The volume of hydrogen evolved is double that of Experiment 1. Since magnesium is the limiting reagent, the mass of magnesium could be doubled.

Mark allocation: 1 mark

- 1 mark for stating the mass of magnesium has doubled.

Note: An increase in acid concentration, or a temperature rise, is not a correct response.

Question 5b.

Worked solution

Possible changes include decreasing the magnesium surface area by adding it as a ball, reducing the volume of the acid and reducing the concentration of acid.

Explanatory note

The total volume of hydrogen gas evolved is the same as that in Experiment 1, but the gas evolves more slowly. This means that the rate of reaction is reduced. This could be caused by conducting the experiment at a lower starting temperature, decreasing the acid concentration or decreasing the surface area of the magnesium.

Mark allocation: 3 marks

- 1 mark for each valid response.

Question 5c.

Worked solution

The change in mass of the reactor over time or the change in pH of the solution over time could be used to monitor the rate of this reaction.

Explanatory notes

$$Mg(s) + 2HCl(aq) \rightarrow MgCl_2(aq) + H_2(g)$$

The evolution of hydrogen gas results in the mass of the beaker or flask decreasing with time. The beaker or flask could be placed on a balance and the mass recorded at regular intervals.

The HCl concentration will also decrease as the reaction proceeds. The pH could be monitored with an appropriate probe.

Mark allocation: 2 marks

- 1 mark for each correct response.

Question 6a.i.

Worked solution

The brown intensity will drop quickly at first and then more slowly. The brown will not completely disappear as bromine in an excess reagent.

Explanatory notes

As the reaction proceeds the bromine concentration drops. When the reaction is complete there is still some bromine present.

Mark allocation: 1 mark

- 1 mark for a description of the change in the intensity of the bromine colour.

Question 6a.ii.

Worked solution

The catalyst will increase the reaction rate. The brown intensity will drop faster but the intensity at the end will be unchanged.

Explanatory notes

Although the catalyst has been added, the amounts have not changed, so the brown intensity at the end will be unchanged.

Mark allocation: 2 marks

- 1 mark for the increased rate.
- 1 mark for stating that the intensity at the end is unchanged.

Question 6b.i.

Worked solution

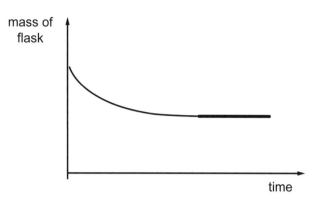

Mark allocation: 1 mark

- 1 mark for the correct graph shape.

Question 6b.ii.

Worked solution

0.8 mol of bromine reacts, therefore 0.8 mol of CO_2 is formed.

mass = $n \times M$ = 0.8 × 44 = 35.2 g

Explanatory notes

The original experiment is between 1.0 mole of bromine and 0.8 mole of methanoic acid. Methanoic acid is the scarce reagent.

Mark allocation: 2 marks

- 1 mark for the correct number of mole of CO_2.
- 1 mark for the correct mass.

Question 6c.

Worked solution

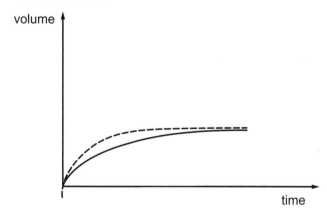

Explanatory notes

The addition of extra bromine will increase the number of particles and the frequency of collisions. However, the amount of methanoic acid has not been increased, so the volume of CO_2 produced will be unchanged.

Mark allocation: 1 mark

- 1 mark for the correct graph.

Question 7a.

Worked solution

$$CH_4(g) + H_2O(g) \rightleftharpoons CO(g) + 3H_2(g)$$

2	1.8	0	0	Start
2 − 0.24	1.8 − 0.24	0.24	0.24 × 3	Equilibrium

$$K = \frac{[CO][H_2]^3}{[CH_4][H_2O]} = \frac{0.24 \times 0.72^3}{1.76 \times 1.56} = 0.033 \text{ M}^2$$

Explanatory notes

The given amounts of methane and steam are not equilibrium values. During the reaction, 0.24 mole of CO forms.

Therefore, the amount of H_2 formed is three times that of CO (i.e. 3 × 0.24 = 0.72 mole).

The amounts of CH_4 and H_2O at equilibrium are 0.24 mole less than their original values. Units for K are required.

Mark allocation: 3 marks

- 1 mark for determining the equilibrium concentrations of the species present.
- 1 mark for the correct equilibrium expression.
- 1 mark for the correct answer and units.

Question 7b.i.

Worked solution

K is unchanged because the temperature is unchanged.

Mark allocation: 1 mark

- 1 mark for stating that K is unchanged.

Question 7b.ii.

Worked solution

The amount of CO will be less because the increase in pressure favours the back reaction.

Mark allocation: 1 mark

- 1 mark for stating that the amount of CO is less.

Question 7b.iii.

Worked solution

The concentration of CO will be greater because the decrease in volume increases the concentration.

Explanatory notes

The concentration of CO has increased even though the back reaction is favoured. This is because of the effect of the volume being halved. When the volume is halved, the pressure is doubled and this outweighs subsequent adjustments.

Mark allocation: 1 mark

- 1 mark for stating that the concentration of CO is increased.

Question 7b.iv.

Worked solution

The rate of the forward reaction will be greater because the decrease in volume will lead to more collisions.

Explanatory notes

The rate of both the forward and back reactions increases equally when the volume is halved. This is due to an increase in the number of collisions.

Mark allocation: 1 mark

- 1 mark for stating that the rate is increased.

Question 7c.

Worked solution

$$CO(g) + 3H_2(g) \rightleftharpoons CH_4(g) + H_2O(g)$$

Explanatory notes

The value of K given in **part c.** is the reciprocal of the original K value. This tells you that the original equation has been reversed.

When the volume is halved, the system will respond by moving in the direction that involves fewer particles. There are two reactant particles compared to four product particles, so the reverse reaction is favoured.

←——————————

$$CH_4(g) + H_2O(g) \rightleftharpoons CO(g) + 3H_2(g)$$

Mark allocation: 1 mark

- 1 mark for supplying the reverse equation of the original equation given.

Question 8a.

Worked solution

Since methanol is the only reactant, the concentration of H_2 gas formed will be double that of CO.

Let $x = [CO]$. Then $[H_2]$ will be $2x$.

$$K = \frac{[CO][H_2]^2}{[CH_3][OH]} = \frac{x(2x)^2}{0.84} = 3.52 \times 10^{-3}$$

$$4x^3 = 0.84 \times 3.52 \times 10^{-3} = 2.96 \times 10^{-3}$$

$$x = \sqrt[3]{7.39} \times 10^{-4} = 0.090$$

Explanatory notes

The key to the question is that methanol is the sole reactant. When it reacts, the amount of H_2 formed will be double that of CO. If x is used to denote the amount of CO, then the amount of H will be $2x$. These symbols can be substituted into the expression for K.

Mark allocation: 4 marks

- 1 mark for recognising that the concentration of H_2 will be double that of CO.
- 1 mark for the correct expression for K.
- 2 marks for correctly calculating the concentration of CO.

Question 8b.i.

Worked solution

Halving the volume doubles the pressure. To oppose the increase in pressure the back reaction is favoured, as the ratio of reactants to products is $1:3$.

Mark allocation: 2 marks

- 1 mark for a comparison of the products and reactants.
- 1 mark for concluding that the back reaction is favoured.

Question 8b.ii.

Worked solution

The concentration will be greater.

Explanatory notes

The concentration of CO is greater after equilibrium is re-established. The change of volume doubles the concentration. The back reaction is favoured but it will only partially oppose the original change.

Mark allocation: 1 mark

- 1 mark for the answer of greater.

Question 8b.iii.

Worked solution

The rate of reaction is greater at the new point of equilibrium because the particles are closer together. The rate of the back reaction has also increased so the value of K is unchanged.

Explanatory notes

The balanced equation $CH_3OH(g) \rightleftharpoons CO(g) + 2H_2(g)$ shows there are two product molecules and only one reactant molecule. To partially oppose an increase in pressure, the reverse reaction is favoured. This reduces the amount of molecules present.

When the volume is halved, the concentration is doubled. The system will move in the reverse direction to re-establish equilibrium, but the concentration of CO will still be higher than before the change.

If the volume is halved, the particles will collide more often. The rate of the forward and back reactions are both increased.

Mark allocation: 1 mark

- 1 mark for stating that the reaction rate is increased.

Question 9a.i.

Worked solution

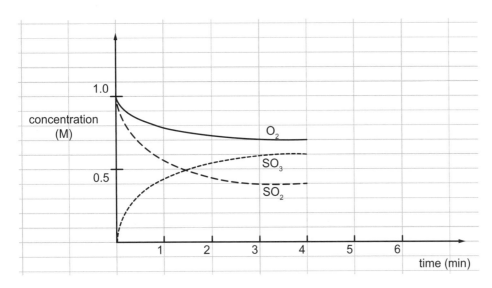

Explanatory notes

The concentration of oxygen drops by 0.3 M. Given the stoichiometry of the reaction, the concentration of SO_2 must drop by 0.6 M and the concentration of SO_3 must increase by 0.6 M. Since the starting amounts of SO_2 and O_2 are equal, the graph for SO_2 should also start at 1.0 M.

Mark allocation: 2 marks

- 1 mark for each consistent concentration (up to 2 marks).

Question 9a.ii.

Worked solution

$$K_c = \frac{[SO_3]^2}{[SO_2]^2[O_2]} = \frac{(0.6)^2}{(0.4)^2(0.7)} = 3.2 \ M^{-1}$$

Explanatory notes

The equilibrium concentrations of all three species are read from the graph and substituted into the expression for K. The graphs drawn need to be consistent with the stoichiometry of the equation for you to obtain the correct answer.

Mark allocation: 2 marks

- 1 mark for substituting concentrations into an expression for K.

- 1 mark for correct answer and units; answer must include one or two significant figures.

TIP

> » Make sure you include a ruler as part of the stationery you bring into the exam. A ruler can be useful to help read graph values correctly or to draw graphs accurately.

> » It is expected that you can determine the correct units for each equilibrium constant calculation.

Question 9b.

Worked solution

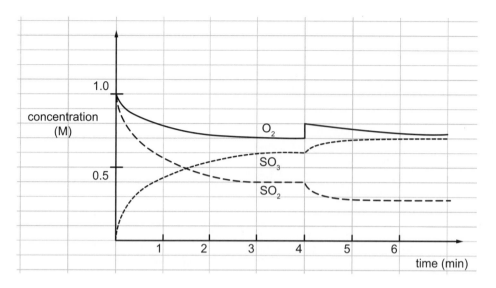

Explanatory notes

The O_2 concentration increases immediately from 0.7 M to 0.8 M. The system opposes this increase by favouring the forward reaction. The concentration of O_2 will drop but it will not drop below 0.7 M.

The concentration of SO_2 will drop by twice as much as the drop in O_2 concentration. The concentration of SO_3 will increase by the same amount as the SO_2 concentration has dropped.

Mark allocation: 3 marks

- 1 mark for each concentration that is consistent with Le Chatelier's principle and the stoichiometry of the equation (up to 3 marks).

Question 9c.i.

Worked solution

The value of K will decrease.

Explanatory notes

This is an exothermic reaction. The value of K will drop when the temperature is increased.

Mark allocation: 1 mark

- 1 mark for stating that the value of K will decrease.

Question 9c.ii.

Worked solution

The amount of SO_3 gas will decrease.

Explanatory notes

If the value of K drops, the back reaction is favoured, therefore lowering the amount of SO_3.

Mark allocation: 1 mark

- 1 mark for stating that the amount of SO_3 will decrease.

Question 9c.iii.

Worked solution

The total amount of gas will increase.

Explanatory notes

If the back reaction is favoured, the amount of reactant increases at the expense of product. The stoichiometry shows that the total amount of gas will increase if the reverse reaction is favoured.

Mark allocation: 1 mark

- 1 mark for stating that the total amount of gas will increase.

Question 10a.

Worked solution

The positive sodium ions will move towards the negative electrode and the negative chloride ions will move towards the positive electrode.

Explanatory notes

Once the NaCl melts, the ions are free to move. Negative ions will move towards the positive electrode (anode) and positive ions will move towards the negative electrode (cathode).

Mark allocation: 2 marks

- 1 mark for stating that the particles are sodium ions and chlorine ions.
- 1 mark for correctly explaining the direction of flow of the ions.

Question 10b.

Worked solution

i. anode: $2Cl^-(l) \rightarrow Cl_2(g) + 2e^-$

ii. cathode: $Na^+(l) + e^- \rightarrow Na(l)$

Explanatory notes

Chloride ions are oxidised to chlorine gas at the anode. Sodium ions form sodium metal at the cathode. Both half-equations can be obtained from the Data Book.

Mark allocation: 2 marks

- 1 mark for the correct half-equation for the anode (states must be correct).
- 1 mark for the correct half-equation at the cathode (states must be correct).

Question 10c.

Worked solution

$2Na^+(l) + 2Cl^-(l) \rightarrow 2Na(l) + Cl_2(g)$

Explanatory notes

The overall equation is obtained by adding together the two half-equations. The sodium half-equation must be multiplied by 2 to balance the electrons.

Mark allocation: 1 mark

- 1 mark for correct equation and states.

Note: Accept $2NACl(l)$.

Question 10d.

Worked solution

Chlorine gas has a pungent odour and is harmful to inhale.

Producing very reactive liquid sodium and chlorine gas in proximity to each other is dangerous, especially at high temperatures.

Explanatory notes

Both elements produced are dangerous. Chlorine is a toxic gas and liquid sodium is very reactive.

Mark allocation: 2 marks

- 1 mark for each valid reason (up to 2 marks).

Question 10e.

Worked solution

$n(e) = 2n(Cl_2) = 2 \times 0.46 = 0.92$ mol

$Q = n(e) \times 96500 = 0.92 \times 96500 = 88800$ C

$t = \dfrac{Q}{i} = \dfrac{88800}{5.62} = 15800$ s (or 4.39 h)

Explanatory notes

The gas produced is chlorine. It takes two electrons to produce one chlorine molecule, therefore the number of mole of chlorine is half the number of mole of electrons.

The charge required is calculated from the formula $Q = n(e) \times 96500$, and time is calculated by rearranging the formula to $Q = It$.

Mark allocation: 3 marks

- 1 mark for the correct number of moles.
- 1 mark for the correct calculation of charge.
- 1 mark for correct calculation of time.

Question 11a.

Worked solution

$3CaO(s) + 2Al(s) \rightarrow Al_2O_3(s) + 3Ca(s)$

Explanatory notes

The equation is straightforward, with aluminium replacing calcium ions. The oxidation number of aluminium will be 3+, so the formula of aluminium oxide is Al_2O_3.

Mark allocation: 1 mark

- 1 mark for a correctly balanced equation.

Note: States can be ignored.

Question 11b.i.

Worked solution

$2Cl^-(l) \rightarrow Cl_2(g) + 2e^-$

Marking allocation: 1 mark

- 1 mark for the correct equation, with states.

Question 11b.ii.

Worked solution

$Ca^{2+}(l) + 2e^- \rightarrow Ca(l)$

Marking allocation: 1 mark

- 1 mark for the correct equation, with states.

Question 11b.iii.

Worked solution

$Ca^{2+}(l) + 2Cl^-(l) \rightarrow Ca(l) + Cl_2(g)$

Explanatory notes

During electrolysis, calcium ions are reduced to calcium metal, and chloride ions are oxidised to chlorine gas. Both half-equations are found in the Data Book. Chloride ions are stronger reductants than fluoride ions, so they react in preference at the anode.

Mark allocation: 1 mark

- 1 mark for the correct equation, with states.

Question 11c.

Worked solution

Charge can be calculated from the number of mole of electrons, and the time can be determined using the formula $t = \dfrac{Q}{I}$.

$$n(Ca) = \frac{1\,000\,000}{40.1} = 24\,900 \text{ mol}$$

$n(e) = 2n(Ca) = 49\,900 \text{ mol}$

$Q = 49\,900 \times 96\,500 = 4.81 \times 10^9 \text{ C}$

$$t = \frac{Q}{I} = \frac{4.81 \times 10^9}{125\,000} = 38\,500 \text{ s} = 10.7 \text{ h}$$

Explanatory notes

1 tonne is 10^6 grams of calcium. The number of mole of calcium needs to be determined and then doubled to give the number of mole of electrons (due to calcium ions being Ca^{2+}). The charge can be calculated from the number of mole of electrons, and the time can be determined using the formula $t = \dfrac{Q}{I}$.

Mark allocation: 3 marks

- 1 mark for correctly calculating $n(e)$.
- 1 mark for correctly calculating the charge.
- 1 mark for correctly calculating the time (which also can be expressed as 10 h 42 min).

Question 11d.

Worked solution

Calcium ions are stronger oxidising agents than potassium ions. The potassium ions will not be reduced and will not interfere with the process.

Explanatory notes

A comparison of the half-equations for calcium and potassium shows that calcium ions will react with chloride ions before potassium ions.

$Cl_2(g) + 2e^- \rightarrow 2Cl^-(l)$

$Ca^{2+}(l) + 2e^- \rightarrow Ca(l)$

$K^+(l) + e^- \rightarrow K(l)$

Mark allocation: 2 marks

- 1 mark for a comparison of the relative oxidising strengths.
- 1 mark for the conclusion that the potassium ions will not be a problem.

Question 12a.

Worked solution

Possible half equations are

$Cl_2(g) + 2e^- \rightleftharpoons 2Cl^-(aq)$

$O_2(g) + 4H^+(aq) + 4e^- \rightleftharpoons 2H_2O(l)$

$Ni^{2+}(aq) + 2e^- \rightleftharpoons Ni(s)$

$2H_2O(l) + 2e^- \rightleftharpoons H_2(g) + 2OH^-(aq)$

Therefore, reactions occurring are

Anode: $2H_2O(l) \rightarrow O_2(g) + 4H^+(aq) + 4e^-$

Cathode: $Ni^{2+}(aq) + 2e^- \rightleftharpoons Ni(s)$

Overall equation: $2H_2O(l) + 2Ni^{2+}(aq) \rightarrow O_2(g) + 4H^+(aq) + 2Ni(s)$

Explanatory notes

There are four relevant half-equations that must be lined up in order of electrochemical series voltage. From this list, the strongest oxidant, Ni^{2+}, will react with the strongest reductant, H_2O.

The reaction of Ni^{2+} to Ni is reduction and it occurs at the cathode. The reaction of water is oxidation and it will occur at the anode.

For the overall equation, the electrons must be balanced. Two nickel ions are needed to balance the production of oxygen gas.

Mark allocation: 4 marks

- 1 mark for evidence of choosing the relevant half-equations.
- 1 mark for the correct anode reaction.
- 1 mark for the correct cathode reaction.
- 1 mark for the correct overall equation.

Question 12b.

Worked solution

Anode: colourless gas forms

Cathode: silver metal forms on the electrode

Explanatory notes

Anode: Oxygen gas and hydrogen ions are produced. An observer will see a colourless gas evolved but will not see any evidence of hydrogen ion formation.

Cathode: Nickel metal is deposited. It will appear as a silver metal.

Mark allocation: 2 marks

- 1 mark for stating that a colourless gas evolves at the anode.
- 1 mark for stating that a grey or silver metal is deposited on the cathode.

Note: No mark for stating 'oxygen gas' is evolved, as an observer could not identify the colourless gas as oxygen.

Question 12c.

Worked solution

$n(\text{Ni}) = \dfrac{2.5}{58.7} = 0.0426 \text{ mol}$

$n(O_2) = \frac{1}{2}\, n(\text{Ni}) = 0.5 \times 0.0426 = 0.0213 \text{ mol}$

$V = n \times 24.5 = 0.0213 \times 24.5 = 0.522 \text{ L}$

Explanatory notes

The number of mole of electrons that pass through the anode is the same as the number that pass through the cathode.

Finding the number of mole of nickel metal deposited can be used to find the number of mole of electrons.

The number of mole of oxygen gas is half the number of mole of nickel metal; that is, $n(O_2) = \frac{1}{2}\, n(\text{Ni})$

The volume of oxygen gas can be determined using $V = n \times 24.5$ because SLC conditions are used.

Mark allocation: 3 marks

- 1 mark for calculating $n(\text{Ni})$.
- 1 mark for calculating $n(O_2)$.
- 1 mark for the correct volume.

Question 13a.i.

Worked solution

The production of hydrogen gas requires significant energy input. Green hydrogen is hydrogen produced using renewable energy.

Mark allocation: 1 mark

- 1 mark for a correct explanation.

Question 13a.ii.

Worked solution

Hydrogen gas can be blended with domestic gas supplies and it can be used in generators and fuel cells.

Mark allocation: 2 marks

- 1 mark for each valid use.

Question 13b.

Worked solution

Anode: $2H_2O(l) \rightarrow O_2(g) + 4H^+(aq) + 4e$

Cathode: $2H^+(aq) + 2e \rightarrow H_2(g)$

Explanatory notes

During electrolysis of water, the water is broken into hydrogen and oxygen gas. Both half-equations required can be found on the electrochemical series. The correct half-equations use H^+ ions due to the acidic conditions used.

Mark allocation: 2 marks

- 1 mark for each correct half-equation.

Question 13c.

Worked solution

Anode: $2H_2O \rightarrow O_2 + 4H^+ + 4e$

Cathode: $2H^+ + 2e \rightarrow H_2$

Explanatory notes

The half-equations occurring will be the same as those for aqueous conditions but the states should not be shown, as a conducting polymer is used instead of an aqueous solution.

Mark allocation: 2 marks

- 1 mark for each correct half-equation.

Question 13d.

Worked solution

Limitation 1: Hydrogen is usually a gas at room temperature. A gas has a lower density, hence it has lower energy density than a liquid. When hydrogen is to be used as a fuel in cars, this is a problem.

Limitation 2: Hydrogen gas is very flammable. It needs to be handled and transported carefully. The need for secure storage adds significantly to its associated costs.

Explanatory notes

To overcome the limitations listed, researchers are considering converting hydrogen gas to a liquid or to ammonia gas. Both are easier to handle and to transport. The conversion to a liquid, however, uses considerable energy.

Mark allocation: 4 marks

- 1 mark for each valid limitation.
- 1 mark for each valid explanation.

Unit 4 | Area of Study 1 How are organic compounds categorised and synthesised?

Multiple choice

Question 1

Answer: B

Explanatory notes

Option B is correct. From left to right, the functional groups present on this molecule are hydroxyl–OH, amine–NH_2 and carboxyl–COOH.

Options A and C are incorrect. They list amide instead of amine.

Option D is incorrect. It lists carbonyl, of which there is none.

Question 2

Answer: C

Explanatory notes

Option C is correct. The general formula of an alcohol is the same as for an alkane, except it has one oxygen atom added.

Options A and B are incorrect. They do not have 2n + 2 for the number of hydrogen atoms.

Option D is incorrect. The number of O atoms is one.

Question 3

Answer: A

Explanatory notes

Option A is correct. The boiling point of propane is lower than that of butane because the molecule is smaller, and the boiling points of the alkanes will be lower than the alcohol's due to the presence of an oxygen atom in the alcohol.

Options B, C and D are incorrect. None list the alcohol as the chemical with the highest boiling point.

Question 4

Answer: D

Explanatory notes

Option D is correct. Numbering for the molecule must start at the right because the hydroxyl functional group has highest priority. The methyl group is a 2-methyl and the alcohol is on the first carbon. The longest carbon chain in this molecule is four carbon atoms, so it is a butane derivative.

Option A is incorrect because numbering of the molecule started at the left.

Option B is incorrect because the methyl group is omitted and the alcohol is listed on the wrong carbon atom.

Option C is incorrect. The numbering has started incorrectly from the left.

Question 5

Answer: **B**

Explanatory notes

Option B is correct. The '2' is necessary in propan-2-ol to describe the position of the hydroxyl group. Butanoic acid does not have the prefix 1- because the position of the carboxyl group is fixed.

Option A is incorrect. It should state simply butanoic acid.

Option C is incorrect because butanoic acid does not have the prefix of 1-.

Option D is incorrect. The first molecule is propan-2-ol.

 TIP

» Note the difference in naming between butanol and butanoic acid. In butanol, the position of the hydroxyl group must be stated (i.e. propan-2-ol), but in butanoic acid the position should not be stated. This is because the position of the hydroxyl group in butanol can vary but the carboxyl group in butanoic acid can only be on the end carbon atom.

Question 6

Answer: **A**

Worked solution

Option A is correct. The molecule should be numbered from the right to give the side chains the lowest possible numbers.

$$CH_3 - CH_2 - CH - CH_2 - CH - CH_3$$

with branches:
CH bearing CH_3 (top), and the other CH bearing $CH_2 - CH_3$ (below).

Numbering for the molecule must start at the right, so the methyl group is on carbon number 2. (Numbering from left to right is 6, 5, 4, 3, 2, 1.)

The ethyl group is on carbon number 4.

The side chains should be in alphabetical order, i.e. ethyl comes before methyl.

Option B is incorrect because numbering starts from the wrong end.

Option C is incorrect because there is a longer chain than butane.

Option D is incorrect. This molecule is not a linear alkane.

 TIP

» It is useful to bring a highlighter into the exam and to trace out the longest possible carbon chain. In this case, the chain will be six carbon atoms long, making this a hexane.

Question 7

Answer: **D**

Explanatory notes

Option D is correct. Butane and the intermediate molecules are drawn below.

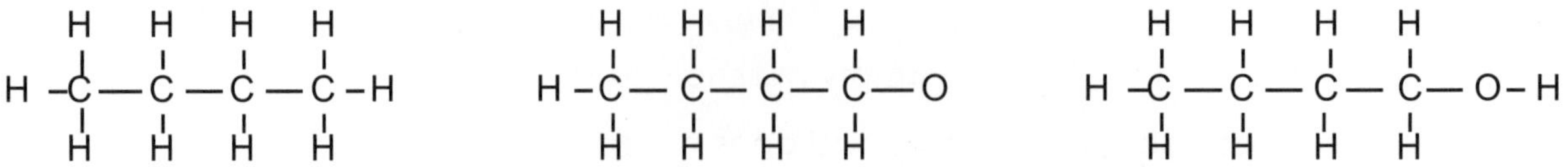

Option A is incorrect. Two substitution reactions would be required, not one.

Option B is incorrect because 1-butanol must be formed before oxidation occurs.

Option C is incorrect because butane is unlikely to undergo addition reactions.

Question 8

Answer: **B**

Explanatory notes

Option B is correct. An amide bond forms from the reaction between a carboxylic acid and an amine.

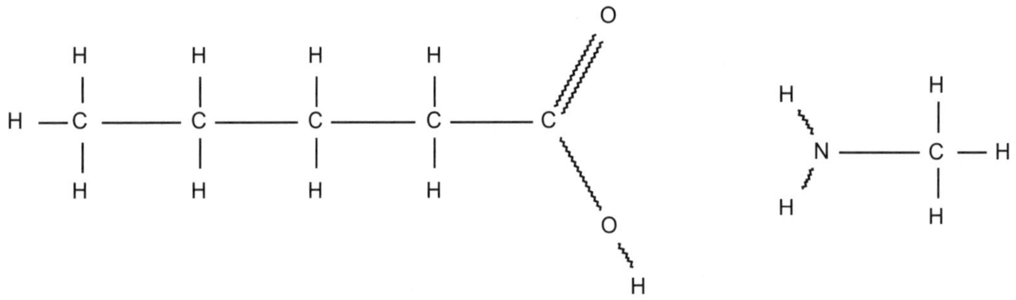

Option A is incorrect. The two functional groups are the wrong way around.

Option C is incorrect because methanamine is required, not ethanamine.

Option D is incorrect. Methanamine is required, not ethanamine. Pentan-1-ol would also need to be pentanoic acid.

Question 9

Answer: **D**

Explanatory notes

Option D is correct. The reaction between a halo compound and ammonia can be used to form an amine.

Option A is incorrect. Neither reactant contains the necessary nitrogen atom.

Options B and C are incorrect because methanamine is not involved.

Question 10

*Answer: **D***

Explanatory notes

Option D is correct. The ester is ethyl ethanoate and this is formed from the reaction between ethanol and ethanoic acid.

Option A is incorrect. Methanoic acid would produce a different ester.

Option B is incorrect becuase the alcohol needs to be ethanol.

Option C is incorrect. Propan-1-ol contains too many carbon atoms for this ester.

 TIP

» Be careful with the semi-structural formulas of esters. The way the ester group is represented depends upon whether the alcohol or the carboxylic acid is drawn first. For example, if the ethanoic acid was drawn first in the ester given in this question, then the semi-structural formula would be $CH_3COOCH_2CH_3$.

Question 11

*Answer: **A***

Explanatory notes

Option A is correct. The substitution of an –OH onto 2-chlorobutane will form propan-2-ol, which can then be oxidised to butan-2-one.

Option B is incorrect. No condensation reaction is required.

Options C and D are incorrect. They are not possible pathways.

Short answer

Question 1a.

Worked solution

Propanol: The name requires a number to show where the hydroxyl group is located. The correct IUPAC name is propan-1-ol.

1-propanoic acid: The 1 in the name is not necessary. The correct IUPAC name is propanoic acid.

Explanatory notes

Numbers are used in names when there are several possibilities for the location of a functional group. Propanol requires a number because there are two possible locations, whereas propanoic acid does not.

Mark allocation: 2 marks

- 1 mark for each correct explanation.

Question 1b.i.

Worked solution

Numbering of carbons must start from the right of this molecule. The correct name is 2-methylpentane.

Explanatory notes

The choice of where to start numbering a molecule must be based on achieving the lowest possible numbers.

Mark allocation: 1 mark

- 1 mark for the correct name.

Question 1b.ii.

Worked solution

The molecule is a structural isomer of hexane rather than of pentane.

Mark allocation: 1 mark

- 1 mark for the correct answer.

Question 1c.i.

Worked solution

from left to right: butan-2-ol and 2-methylpropan-2-ol

Mark allocation: 2 marks

- 1 mark for each correct answer.

Question 1c.ii.

Worked solution

In a secondary alcohol, the –OH is connected to a carbon that in turn is connected to two other carbon atoms.

In a tertiary alcohol, the –OH is connected to a carbon that in turn is connected to three other carbon atoms.

Mark allocation: 2 marks

- 1 mark for each type of alcohol explained correctly.

Question 2a.i.

Worked solution

Both molecules are non-polar because they do not contain any highly electronegative atoms.

Explanatory notes

The alkanes are non-polar molecules that can be used as fuels. They have weak dispersion forces between molecules.

Mark allocation: 2 marks

- 1 mark for the valid property.
- 1 mark for a correct explanation.

Question 2a.ii.

Worked solution

The boiling point of hexane is significantly higher than ethane due to the stronger dispersion forces between molecules.

Note: Student responses could refer to other properties, such as viscosity.

Mark allocation: 2 marks

- 1 mark for a valid property.
- 1 mark for a correct explanation.

Question 2b.

Worked solution

i. molecular formula: $C_2H_5O_2$

ii. structural diagram:

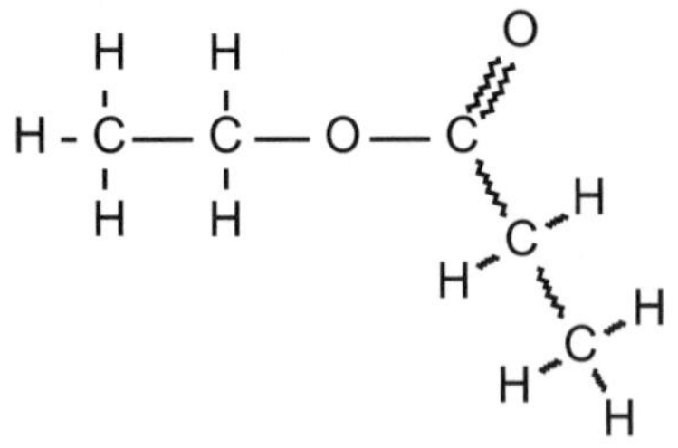

iii. semi-structural formula: $CH_3CH_2OCOCH_2CH_3$

iv. skeletal structure:

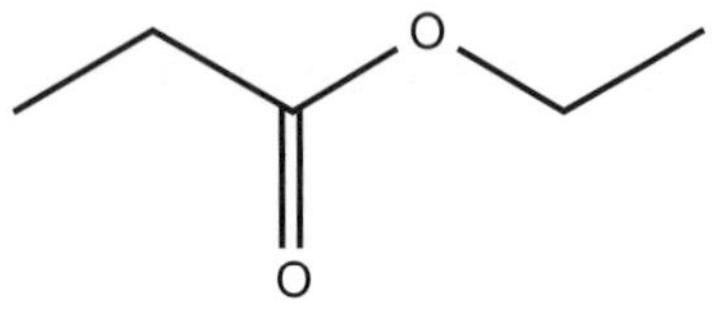

Mark allocation: 4 marks

- 1 mark for each correct representation.

Question 3a.

Worked solution

i. $C_6H_{12}O_2$

ii. C_3H_6O

Explanatory notes

The molecular formula can be determined by adding the atoms of each element. Alternatively, saturated esters will have twice as many hydrogen atoms as carbon atoms.

The molecular formula can be halved to obtain the empirical formula.

Mark allocation: 2 marks

- 1 mark for the correct molecular formula.
- 1 mark for the correct empirical formula.

Question 3b.i.

Worked solution

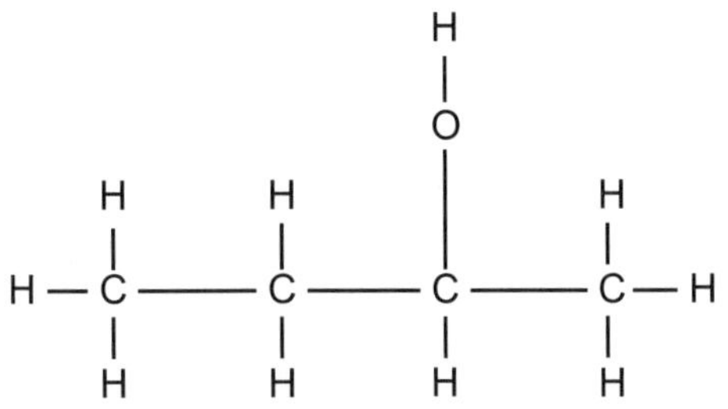

Mark allocation: 1 mark

- 1 mark for the structure of butan-2-ol drawn correctly.

Question 3b.ii.

Worked solution

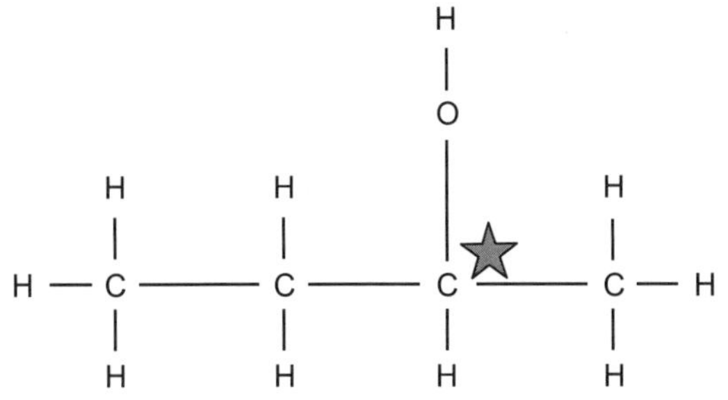

Mark allocation: 1 mark

- 1 mark for a star positioned on the correct carbon.

Question 3b.iii.

Worked solution

butan-2-ol

Mark allocation: 1 mark

- 1 mark for butan-2-ol.

Note: Butanol is not an acceptable answer. 2-butanol is acceptable but is not the preferred representation.

Question 3b.iv.

Worked solution

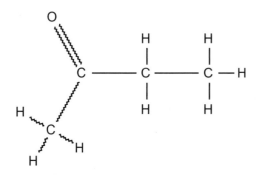

Explanatory notes

Hydrolysis of an ester will produce an alcohol and a carboxylic acid. The alcohol is butan-2-ol. This is a secondary alcohol that can be oxidised to a ketone.

Mark allocation: 1 mark

- 1 mark for the correct structure.

Question 3c.i.

Worked solution

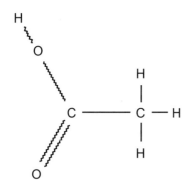

Explanatory notes

The monoprotic carboxylic acid formed contains two carbon atoms, therefore it will be ethanoic acid.

Mark allocation: 1 mark

- 1 mark for drawing ethanoic acid correctly.

Question 3c.ii.

Worked solution

Explanatory notes

The reaction between a carboxylic acid and an amine is a condensation reaction that forms the amide drawn, as well as water.

Mark allocation: 2 marks

- 1 mark for drawing the amide shown.
- 1 mark for drawing water.

Question 4a.i.

Worked solution

An obvious molecule is chloroethane:

Explanatory notes

Alcohols can be formed by reaction of chloroalkanes with KOH or NaOH. The hydroxyl group is substituted for the chlorine atom. A salt is also formed.

Chloroethane is the obvious example, but other halogens could be used in place of chlorine.

Mark allocation: 1 mark

- 1 mark for drawing chloroethane or a similar haloalkane.

> **TIP**
>
> » Organic flow chart questions are very common. You should rote learn the reagents or catalysts required for each step. Read the question carefully to see whether structural, skeletal or semi-structural responses are required.

Question 4a.ii.

Worked solution

The formula for the reaction with chloroethane is

$C_2H_5Cl(g) + KOH(aq) \rightarrow C_2H_5OH(aq) + KCl(aq)$

Explanatory notes

The hydroxyl group is substituted for the chlorine atom. The salt formed is KCl. NaOH could be substituted for KOH or OH− without the spectator ion.

Mark allocation: 1 mark

- 1 mark for a correctly balanced equation; states are not required.

Question 4b.i.

Worked solution

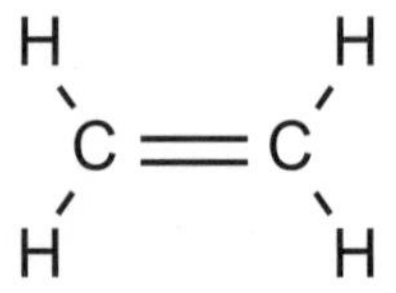

Explanatory notes

Ethene can be used to produce ethanol in an addition reaction.

Mark allocation: 1 mark

- 1 mark for drawing ethene.

Question 4b.ii.

Worked solution

$C_2H_4(g) + H_2O(g) \rightarrow C_2H_5OH(g)$

Explanatory notes

There is only one product, ethanol, in this reaction. The carbon-to-carbon double bond breaks when hydrogen atoms and hydroxyl groups bond to the molecule.

Mark allocation: 1 mark

- 1 mark for a correctly balanced equation; states are not required.

Question 4c.i.

Worked solution

Explanatory notes

Glucose can be converted by yeast to ethanol in fermentation reactions.

Mark allocation: 1 mark

- 1 mark for a sketch of glucose.

Note: The sketch could match the one provided in the Data Book or it could show all bonds.

Question 4c.ii.

Worked solution

$C_6H_{12}O_6(aq) \rightarrow 2C_2H_5OH(aq) + 2CO_2(g)$

Explanatory notes

Glucose reacts in anerobic conditions to form ethanol and CO_2.

Mark allocation: 1 mark

- 1 mark for a correctly balanced equation.

Note: States are not required, but the mark is forfeited if ethanol is shown as a liquid.

Question 4d.

Worked solution

The addition reaction involving ethene will have the highest atom economy because there is only one product formed.

Explanatory notes

Addition reactions lead to one product only. If there is one product only, the atom economy must be 100%.

Mark allocation: 2 marks

- 1 mark for identifying the addition reaction.
- 1 mark for a justification referring to addition reactions.

Question 5a.

Worked solutions

A: propene

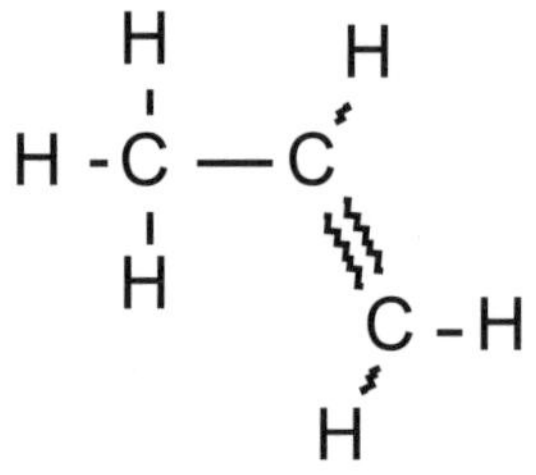

H H
| |
H -C — C
| \\\
H C -H
|
H

B: propan-1-ol

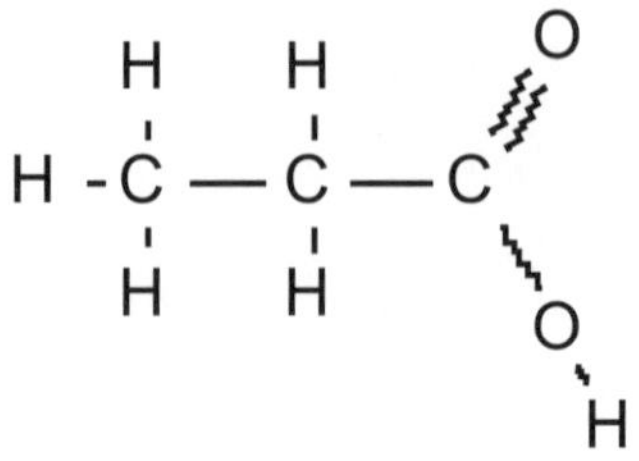

H H O
| | //
H -C — C — C
| | \
H H O
\
H

C: propanoic acid

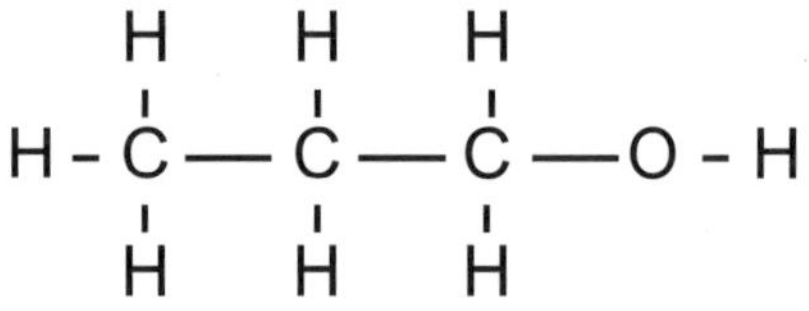

H H H
| | |
H - C — C — C —O - H
| | |
H H H

D: ethanamine

H H H
| | /
H -C — C — N
| | \
H H H

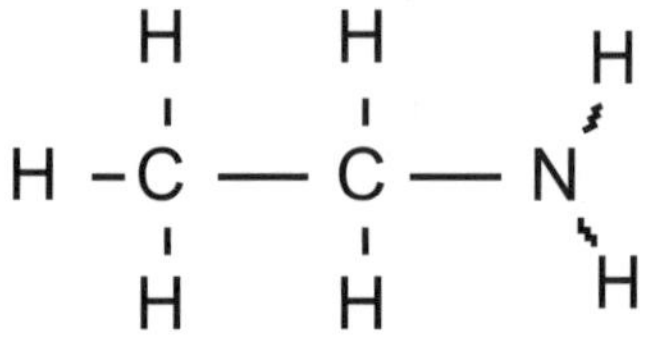

Marking allocation: 8 marks

- 1 mark for each correct structure.
- 1 mark for each correct name.

Question 5b.

Worked solution

$Cr_2O_7^{2-}(aq) + 14H^+(aq) + 6e^- \rightarrow 2Cr^{3+}(aq) + 7H_2O(l)$

Marking allocation: 1 mark

- 1 mark for a correct half-equation.

Unit 4 | Area of Study 2 How are organic compounds analysed and used?

Multiple choice

Question 1

Answer: A

Explanatory notes

Option A is correct. The molecule should contain a carbon-to-carbon double bond if it reacts with bromine, and it should contain a carboxyl group to react with NaOH. It will not react with $Cr_2O_7^{2-}$(aq) because the COOH group is already fully oxidised. Molecule A has all these features.

Option B is incorrect. It does not contain a carbon-to-carbon double bond.

Option C is incorrect. It contains neither functional group.

Option D is incorrect. It does not contain a carboxyl group.

Question 2

Answer: B

Explanatory notes

Option B is correct. The results are all consistent, therefore they are precise. They do not match the tested value, however, so they are not accurate.

Option A is incorrect. The results are precise.

Option C is incorrect. The results do not match the tested value, so they are not accurate.

Option D is incorrect. The results are not accurate.

Question 3

Answer: D

Explanatory notes

Option D is correct. The extra water dilutes the potassium permanganate solution, hence a larger titre is required to achieve the same number of mole of solution. This will lead to a high estimate of the concentration of the iron solution.

Option A is incorrect. A burette should be rinsed with the solution that is to go into it rather than with water, so as not to dilute the solution.

Option B is incorrect. The titre will be high.

Option C is incorrect. The titre will be high.

Question 4

Answer: A

Explanatory notes

Option A is correct.

$n(MnO_4^-) = c \times V = 0.26 \times 0.02 = 0.0052$ mol

$n(Fe^{2+}) = 5 \times 0.0052 = 0.026$ mol

$c(Fe^{2+}) = \dfrac{n}{V} = \dfrac{0.026}{0.025} = 1.04$ M

Option B is incorrect. A mole ratio of 2 : 1 is not correct.

Option C is incorrect. The mole ratio is not 1 : 1.

Option D is incorrect. The mole ratio has been applied in reverse.

Question 5

Answer: A

Explanatory notes

Option A is correct. Oxidation of propan-1-ol produces propanoic acid. This is a weak acid with a pH around 3.

Option B is incorrect. It would produce propanone, which is not acidic.

Options C and D are incorrect. They are unlikely to react with dichromate ions.

Question 6

Answer: B

Explanatory notes

Option B is correct. There is no reason for the melting point of a substance to be above the melting point of the pure sample. If the sample has impurities, the melting point will be lower rather than higher.

Options A, C and D are incorrect. They are not consistent with a sample of high purity or impure salicylic acid.

Question 7

Answer: A

Explanatory notes

Option A is correct. A protein such as collagen is the only alternative that will have a significant number of nitrogen atoms.

Option B is incorrect. A triglyceride will not contain as many oxygen or nitrogen atoms.

Options C and D are incorrect. Polysaccharides are unlikely to have several nitrogen atoms.

Question 8

*Answer: **D***

Explanatory notes

Option D is correct. The six combinations are Ile-Leu-Lys, Ile-Lys-Leu, Leu-Lys-Ile, Leu-Ile-Lys, Lys-Ile-Leu and Lys-Leu-Ile.

Option A is incorrect. There are six possibilities.

Option B is incorrect. Keep in mind that the order of the amino acids is significant.

Option C is incorrect. There are six possibilities.

Question 9

*Answer: **A***

Explanatory notes

Option A is correct. Triglycerides contain ester linkages. Hydrolysis breaks the ester bonds to form carboxyl groups and hydroxyl groups.

Option B is incorrect. Hydrolysis of proteins produces amine and carboxyl groups.

Option C is incorrect. Hydrolysis of carbohydrates produces hydroxyl groups.

Option D is incorrect. The products of hydrolysis of vitamins are varied and complex.

Question 10

*Answer: **A***

Explanatory notes

Option A is correct. Bile emulsifies large globules of fat to smaller globules. Lipase can then hydrolyse the triglycerides to fatty acids and glycerol.

Option B is incorrect. Lipase does not play a role in transport.

Option C is incorrect. The action of lipase is specific to triglycerides, not proteins. The name lipase means cutting lipids.

Option D is incorrect. It is bile that leads to the formation of emulsions.

Question 11

Answer: **C**

Explanatory notes

Option C is correct. If the empirical formula is $C_9H_{16}O$, then the molecular formula will be $C_{18}H_{32}O_2$. This has two carbon-to-carbon double bonds. A saturated fatty acid will have a general formula of $C_nH_{2n}O_2$. The fatty acid in this question is four hydrogen atoms short of saturation. The introduction of a C=C bond causes the number of hydrogen atoms to drop by two, therefore there must be two C=C double bonds.

Option A is incorrect. This is not a saturated molecule.

Option B is incorrect. The question lists an empirical formula. Once this formula is doubled to give the molecular formula, the correct number of carbon-to-carbon double bonds can be determined.

Option D is incorrect. This molecule has two carbon-to-carbon double bonds, not three.

TIP

» You need to have a quick way of determining the number of carbon-to-carbon double bonds in a fatty acid. The general formula of a saturated fatty acid is $C_nH_{2n+1}COOH$ or $C_nH_{2n}O_2$. Each carbon-to-carbon double bond causes the number of hydrogen atoms present to drop by 2.

Question 12

Answer: **D**

Explanatory notes

Option D is correct. The –O-H (acid) absorption band will be around 3000 cm^{-1}. Propanal and propan-1-ol will not have this absorption.

Option A is incorrect. Propanal would also have this absorption.

Option B is incorrect. All three molecules will have a C–H absorption at 3000 cm^{-1}.

Option C is incorrect. A broad peak at 3300 cm^{-1} would suggest propan-1-ol.

Question 13

Answer: **D**

Explanatory notes

Option D is correct. From the structure of 2-chloropropane, it can be seen that the middle proton has six neighbouring protons. These protons have the same environment. Using the n + 1 rule, the six protons will produce a septet.

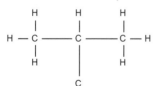

Option A is incorrect. It should be a septet.

Option B is incorrect. It should be a septet.

Option C is incorrect. It should be a septet. (Six identical protons will produce n + 1 (6 + 1) peaks.)

Question 14

*Answer: **D***

Explanatory notes

Option D is correct. The molecule must have three different carbon environments, therefore ruling out propanone and propan-2-ol. The shifts shown match propan-1-ol (10 ppm = $R–CH_3$, 25 ppm = $R–CH_2–R$ and 65 ppm = $R–CH_2OH$) rather than propanoic acid.

Option A is incorrect. Propan-2-ol contains only two carbon environments.

Option B is incorrect. Propanone contains only two carbon environments.

Option C is incorrect. The shifts do not match that of propanoic acid.

Question 15

*Answer: **D***

Explanatory notes

Option D is correct. Cyclohexane is the only option that has only one carbon environment (all $–CH_2–$ groups within the ring are equivalent). The shift will be a low one, matching the spectrum shown, because cyclohexane does not contain highly electronegative atoms.

Option A is incorrect. Propane has two carbon environments.

Option B is incorrect. Butane has two carbon environments.

Option C is incorrect. Cyclohexene has three carbon environments.

TIP

» **Cyclohexane and benzene are in the Study Design. Be aware of their structures and properties.**

Question 16

*Answer: **C***

Explanatory notes

Option C is correct. The three hydrogen atoms on each end are equivalent. There are two neighbouring hydrogen atoms, so the peak is split into $n + 1 = 3$, a triplet. The two middle hydrogen atoms are equivalent. They have six neighbouring hydrogen atoms, so the peak is split into a septet.

Option A is incorrect. The hydrogen atoms are not equivalent.

Option B is incorrect. Propane will not have a quartet.

Option D is incorrect. Propane has two hydrogen environments.

Question 17

Answer: **C**

Explanatory notes

Option C is correct. The retention time can be used to identify butan-1-ol as the alcohol and the area is one-third of the original. The concentration will therefore be one-third.

Options A and B are incorrect. They refer to propan-1-ol.

Option D is incorrect. It has the concentration the wrong way around.

Question 18

Answer: **B**

Explanatory notes

Option B is correct. Since it is a sample of petrol, most components should be hydrocarbons. The shorter molecules will have the lower retention times because the solvent is a non-polar one.

Option A is incorrect. There might be two molecules present with the same retention time, or components that do not travel through the column.

Option C is incorrect. Low molecular mass molecules will have the shortest retention time.

Option D is incorrect. The response of one compound is not necessarily the same as that of another compound.

Question 19

Answer: **B**

Explanatory notes

Option B is correct. As the number of hydroxyl groups increases, the polarity of the molecule will increase. This highly polar molecule will adsorb strongly to the stationary phase, leading to a high retention time.

Option A is incorrect. A highly polar molecule will have a relatively long retention time in a polar stationary phase.

Option C is incorrect. The increase in polarity will be more significant than the increase in relative molecular mass.

Option D is incorrect. A polar molecule will not be highly soluble in a non-polar solvent such as hexane.

 TIP

» Students' performance on questions involving high performance liquid chromatography (HPLC) has not been strong over the past few years. Make sure you understand the significance of the choices of stationary phases and mobile phases, and the types of bonding involved.

Question 20

*Answer: **D***

Explanatory notes

Option D is correct. Run 2 shows two peaks, hence it features two alcohols. The areas under those two peaks in Run 2 are double and half, respectively, of Run 1.

Option A is incorrect. The retention times have not changed.

Options B and C are incorrect. There is no evidence that the experimental set-up has changed.

Question 21

*Answer: **B***

Explanatory notes

Option B is correct. The fastest rate of reaction is when the time taken is the shortest time, which is at 40 °C.

Option A is incorrect. The reaction is much faster at 40 °C.

Option C is incorrect. The reaction is much faster at 40 °C.

Option D is incorrect. The long time taken for the reaction shows that the enzyme is barely functioning at this temperature.

Question 22

*Answer: **A***

Explanatory notes

Option A is correct. Benedict's solution detects the presence of reducing sugars (such as monosaccharides). The glucose formed from the hydrolysis of sucrose is a reducing sugar.

Option B is incorrect. Sucrose will not react with Benedict's solution.

Option C is incorrect. Invertase is an enzyme and not a sugar.

Option D is incorrect. The Benedict's solution does not contain protein.

Question 23

*Answer: **B***

Explanatory notes

Option B is correct. Bread is high in carbohydrates. The amylase hydrolyses some of the starch to form smaller, sweeter sugars.

Option A is incorrect. Chewing breaks up the food but it does not break the molecules themselves.

Option C is incorrect. Saliva will not act on proteins.

Option D is incorrect. It is not relevant.

Question 24

Answer: **C**

Explanatory notes

Option C is correct. Ionic bonds are part of the tertiary structure of a protein. They form between side chains of some amino acids.

Option A is incorrect. This is an ionic bond, not a hydrogen bond.

Option B is incorrect. The bond shown is an ionic bond.

Option D is incorrect. Ionic bonds are not part of the primary structure of a protein.

Question 25

Answer: **A**

Explanatory notes

Option A is correct. Liver contains an enzyme that catalyses this reaction. The effect of high temperature on an enzyme is to denature it, making it ineffective.

Option B is incorrect. Grinding the catalyst increases the surface area and consequently the reaction rate.

Option C is incorrect. A higher temperature will increase the rate of reaction when an inorganic catalyst is used.

Option D is incorrect. Mincing the liver will increase its surface area and its effectiveness.

Short answer

Question 1a.

Worked solution

$5C_2O_4H_2(aq) + 2MnO_4^-(aq) + 6H^+(aq) \rightarrow 2Mn^{2+}(aq) + 10CO_2(g) + 8H_2O(l)$

Explanatory notes

The two half-equations are:

$C_2O_4H_2(aq) \rightarrow 2CO_2(g) + 2H^+(aq) + 2e$

$MnO_4^-(aq) + 8H^+(aq) + 5e \rightarrow Mn^{2+}(aq) + 4H_2O(l)$

When the half-equations are balanced and combined, the overall equation above is obtained.

Mark allocation: 2 marks

- 1 mark for the correct reactants and products.
- 1 mark for the correctly balanced equation.

Question 1b.i.

Worked solution

$n(MnO^{4-}) = c \times V = 0.12 \times 0.02 = 0.00240$ mol

Explanatory notes

The concentration and volume of MnO^{4-} are both known, so the number of moles can be calculated as the starting point.

Mark allocation: 1 mark

- 1 mark for the correct calculation of moles.

Question 1b.ii.

Worked solution

$n(\text{oxalic acid}) = \dfrac{5}{2n}(MnO^{4-}) = \dfrac{5}{2} \times 0.0024 = 0.006$ mol

Explanatory notes

The number of moles of oxalic acid will be $\dfrac{5}{2}$ times the number of moles of MnO^{4-}, from the balanced equation.

Mark allocation: 1 mark

- 1 mark for the correct number of moles.

Note: Consequential marking will apply with this question. If you did not balance the equation correctly, you can still obtain full marks for subsequent questions.

Question 1b.iii.

Worked solution

n(oxalic acid) in the volumetric flask = n(titration) $\times \dfrac{250}{14.8}$

$$= 0.006 \times \dfrac{250}{14.8} = 0.101 \text{ mol}$$

m(oxalic acid) = $n \times M$ = 0.101 × 90.0 = 9.12 g

Explanatory notes

The number of moles of oxalic acid in the volumetric flask is greater than the number of moles in the titre. The ratio of volumes is 250 : 14.8.

Mark allocation: 2 marks

- 1 mark for applying the dilution factor correctly.
- 1 mark for correctly calculating the mass.

Note: There are alternative ways of reaching the correct answer. Award 2 marks for the correct answers if the working is valid.

Question 1b.iv.

Worked solution

The % (m/m) of oxalic acid = $\dfrac{9.12 \times 100}{10.0}$ = 91.2%

Mark allocation: 1 mark

- 1 mark for the correct calculation.

Question 2

Worked solution

Test	Observation	Explanation
Acidified $Cr_2O_7^{2-}$ is heated with ethanol.	The orange colour changes to green.	The dichromate oxidises the ethanol to ethanoic acid. The orange colour of dichromate is replaced by the green colour of the Cr^{3+} ions.
Acidified $Cr_2O_7^{2-}$ is heated with 2-methylpropan-2-ol.	No change occurs.	2-methylpropan-2-ol is a tertiary alcohol. Tertiary alcohols do not react with dichromate.
Iodine solution is added from a burette to linolenic acid.	The brown colour initially disappears then eventually re-appears and lingers.	The iodine reacts with the C=C double bonds in linolenic acid. Once the double bonds have reacted, the reaction stops.

Mark allocation: 6 marks

- 1 mark for each correct answer (up to 6 marks).

Question 3

Worked solution

Macronutrient	Draw the bond broken during hydrolysis	Name of product(s) of hydrolysis
protein	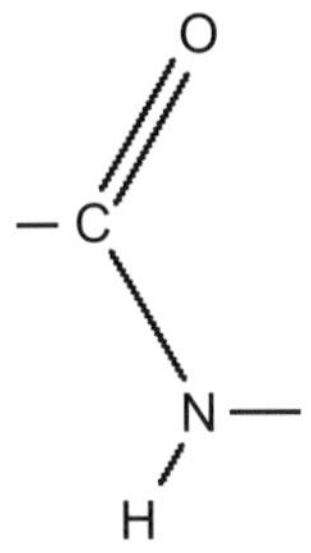	amino acids
carbohydrate	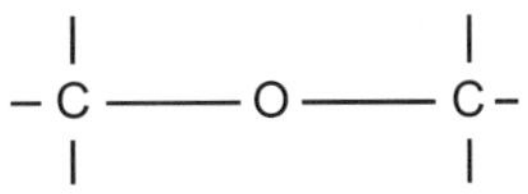	glucose
triglyceride	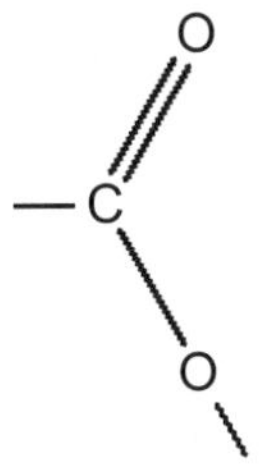	glycerol and fatty acids

Explanatory notes

Proteins: The amide bond is broken to form the amino acids that proteins are built from.

Carbohydrates: The glycosidic bonds are broken to form monosaccharide molecules.

Triglycerides: The ester bonds are broken to form glycerol and fatty acids.

Mark allocation: 6 marks

• 1 mark for each cell filled in correctly (up to 6 marks).

Question 4a.i.

Worked solution

$CH_3(CH_2)_4(CH=CHCH_2)_2(CH_2)_6COOH$ or $CH_3(CH_2)_4CHCHCH_2CHCH (CH_2)_7COOH$

Explanatory notes

The fatty acid is formed from the hydrolysis of the ester. The semi-structural formula can be established by working carefully from the left of the molecule.

Mark allocation: 1 mark

• 1 mark for the correct semi-structural diagram.

Question 4a.ii.

Worked solution

linoleic acid

Explanatory notes

Inspection of the Data Book confirms it is linoleic acid.

Mark allocation: 1 mark

- 1 mark for linoleic acid.

Question 4b

Worked solution

$$CH_3(CH_2)_4(CH=CHCH_2)_2(CH_2)_6COOCH_2$$
$$|$$
$$CH_3(CH_2)_4(CH=CHCH_2)_2(CH_2)_6COOCH$$
$$|$$
$$CH_3(CH_2)_4(CH=CHCH_2)_2(CH_2)_6COOCH_2$$

Explanatory notes

Triglycerides are formed when three fatty acid molecules combine with glycerol. Three ester bonds are formed when this occurs.

Mark allocation: 2 marks

- 1 mark for three fatty acids attached to one glycerol molecule.
- 1 mark for the correct ester bond representation.

Note: Triglycerides are difficult to draw. Accept some combinations of structural and semi-structural formats, but the ester must be correct and the presence of three fatty acid chains evident.

Question 5a.i.

Worked solution

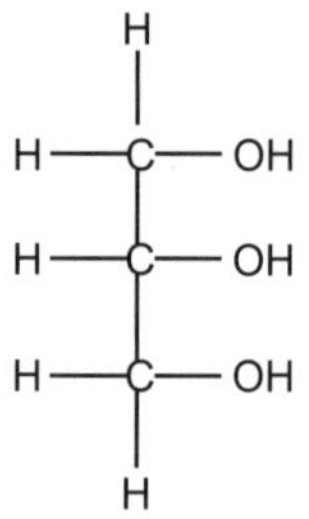

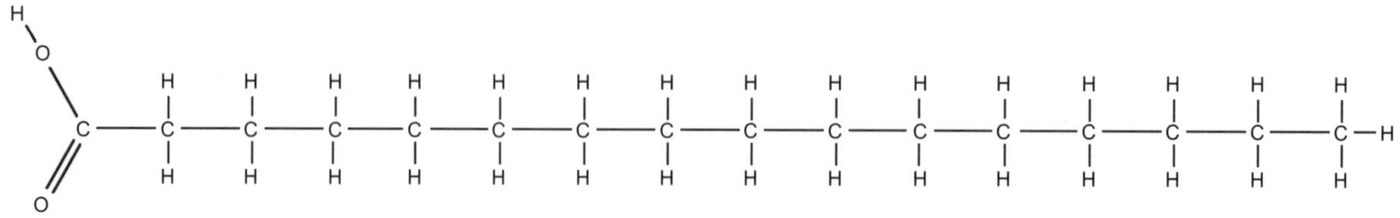

Mark allocation: 2 marks

- 1 mark for the structure of glycerol.
- 1 mark for the structure of the fatty acid.

Question 5a.ii.

Worked solution

Most digestion of fats and oils occurs in the small intestine. Fats and oils are non-polar, so do not break down to any great extent in the stomach. The presence of emulsifiers in the small intestine leads to the breaking down of fats.

Explanatory notes

Triglycerides are formed from a reaction between glycerol and fatty acids. Hydrolysis reverses this reaction, re-forming the glycerol and fatty acids. The fatty acids in this particular triglyceride are all the same.

Mark allocation: 2 marks

- 1 mark for large intestine.
- 1 mark for mentioning the non-polar nature of fats and oils.

Question 5b.i.

Worked solution

$CH_3(CH_2)_{14}COOCH_3$

Explanatory notes

Biodiesel is an ester, in this case between methanol and the fatty acid drawn earlier. The fatty acid has 16 carbon atoms and the methanol adds one more.

Mark allocation: 1 mark

- 1 mark for the semi-structural formula of biodiesel.

Question 5b.ii.

Worked solution

$2C_{17}H_{34}O_2(l) + 49O_2(g) \rightarrow 34CO_2(g) + 34H_2O(g)$

Explanatory notes

Complete combustion of the biodiesel will form CO_2 and H_2O. Be systematic: balance the carbon atoms first, then the hydrogen atoms and finally the oxygen atoms.

Mark allocation: 2 marks

- 1 mark for the correct formulas of the reactants.
- 1 mark for a correctly balanced equation with correct states.

Question 6

Worked solutions

a. B

b. D

c. F

d. G

e. C

f. E

Explanatory notes

a. $C_{19}H_{38}O_2$, represents stearic acid bonded to methanol, a typical biodiesel ester.

b. Glucose has the formula $C_6H_{12}O_6$. When a disaccharide forms, this formula will be duplicated but water is eliminated. This leads to $C_{12}H_{22}O_{11}$.

c. A by-product of the hydrolysis of lipids is glycerol. The molecular formula of glycerol is $C_3H_8O_3$.

d. $C_{18}H_{34}O_2$ is an unsaturated fatty acid. If it was saturated, the formula would be $C_{18}H_{36}O_2$.

e. $C_6H_{12}O_6$ is glucose, produced by photosynthesis.

f. Proteins are formed from amino acids. The amino acid is molecule E, $C_3H_7O_2NS$. This is cysteine.

Mark allocation: 6 marks

- 1 mark for each correct response.

 TIP

» Many clues to help answer questions of this nature can be found in the Data Book. The molecular formulas of the fatty acids are provided and the structures of simple sugars and bases are shown in the Data Book.

Question 7

Worked solution

Chemical	State
glucose in fermentation	(aq)
sulfuric acid as an esterification catalyst	(l)
H_2O reacting with ethene to form ethanol	(g)
biodiesel in combustion	(l)
potassium metal produced in electrolysis	(l)

Explanatory notes

Fermentation to form alcohol occurs in an aqueous environment. Yeast is a living organism that cannot survive when the alcohol concentration is too high.

Esterification requires that the carboxylic acid, alcohol and sulfuric acid are all in the liquid state and not aqueous.

The reaction of ethene and water requires high temperatures, where the water will be present as steam.

Water inhibits the combustion of fuels, especially non-polar fuels such as biodiesel.

Potassium metal is obtained by electrolysis of molten potassium salts. The metal is produced as a liquid.

Mark allocation: 5 marks

• 1 mark for each correct response (up to 5 marks).

TIP

» It expected that students include the correct states when writing balanced equations. The five examples included in this question have all appeared on recent VCAA Chemistry exams.

Question 8a.

Worked solution

prop-2-en-1-ol   propan-2-one   propanal

Explanatory notes

Three possibilities are provided above. The number of hydrogen atoms makes it unlikely that the molecule is an alcohol, unless it also has a carbon-to-carbon double bond.

Mark allocation: 3 marks

- 1 mark for each correct structure and correct name (up to 3 marks).

Note: There might be other possible answers that match this molecular formula.

Question 8b.i.

Worked solution

m/z ratio = 29

Mark allocation: 1 mark

- 1 mark for the correct answer.

Question 8b.ii.

Worked solution

$CH_3CH_2^+$ and CHO^+

Explanatory notes

The fragment must have a mass of 29 g/mol and be part of the molecular formula C_3H_6O.

Mark allocation: 2 marks

- 1 mark for each correct answer (up to 2 marks).

Note: An answer must have a positive charge to receive a mark. $CH_3CH_2^+$ can be shown as $C_2H_5^+$.

Question 8c.i.

Worked solution

No, as there is no broad peak around 3300 cm^{-1}.

Explanatory notes

An –O-H group will show on infrared spectra as a broad peak around 3300 cm^{-1}. This spectrum does not have this peak.

Mark allocation: 1 mark

- 1 mark for the correct answer with an appropriate explanation.

Question 8c.ii.

Worked solution

Yes, as there is strong absorption around 1700 cm^{-1}.

Explanatory notes

A C=O group will show on infrared spectra as a peak around 1700 cm^{-1}. This spectrum has this peak.

Mark allocation: 1 mark

- 1 mark for the correct answer with an appropriate explanation.

Question 8d.

Worked solution

The molecule is propanal.

Propanal should have three different hydrogen environments, a quintet and two triplets.

The shift of 9.8 matches that of an aldehyde (–CHO).

Explanatory notes

Propanal has three different hydrogen environments. The –CHO peak will have a shift of 9.8 (from the Data Book) and it will be a triplet because it has two neighbouring protons. The middle carbon has four neighbouring hydrogen atoms, hence it will be split into five. (It is actually an uneven split becuase the neighbouring protons are not equivalent, but this is beyond the scope of the VCE course). The methyl group on the end will be split into a triplet because it has two neighbouring protons.

Mark allocation: 2 marks

- 1 mark for the correct choice of propanal.
- 1 mark for valid explanation that refers to the hydrogen environments.

Question 9a.

Worked solution

Propan-1-ol produced the spectrum. Both molecules will have a parent molecular ion with a m/z ratio of 60, so this cannot be used to distinguish the molecules. The peak at 31 is consistent with the $-CH_2OH^+$ fragment of a primary alcohol and the peak at 29 is consistent with an ethyl group. Additionally, the peak at 43 is consistent with the $-CH_3CH_2CH_2^+$ fragment. Propan-1-ol contains all these features.

Ethanoic acid would likely have a bigger peak at 15, matching the methyl group, and a large peak at 45 matching the $-COOH^+$ fragment. These peaks are not evident.

Explanatory notes

A characteristic fragmentation of a primary alcohol forms the $-CH_2OH^+$ fragment, which is shown in the diagram below.

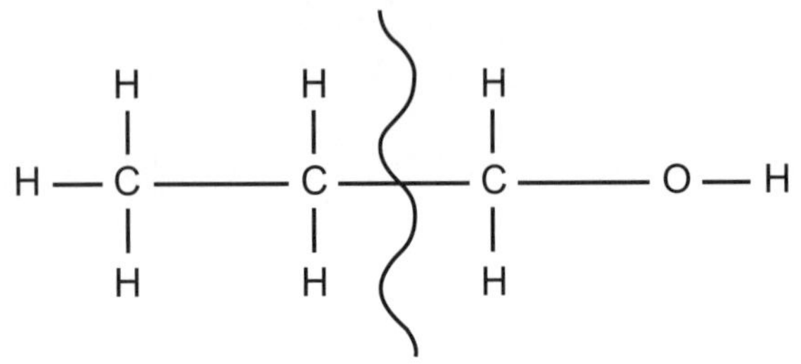

This will also form a peak with a m/z ratio of 29 due to the $CH_3CH_2^+$.

Mark allocation: 3 marks

- 1 mark for nominating propan-1-ol as the correct structure.
- 1 mark for valid reasons why it is propan-1-ol.
- 1 mark for valid reasons why it is not ethanoic acid.

Question 9b.

Worked solution

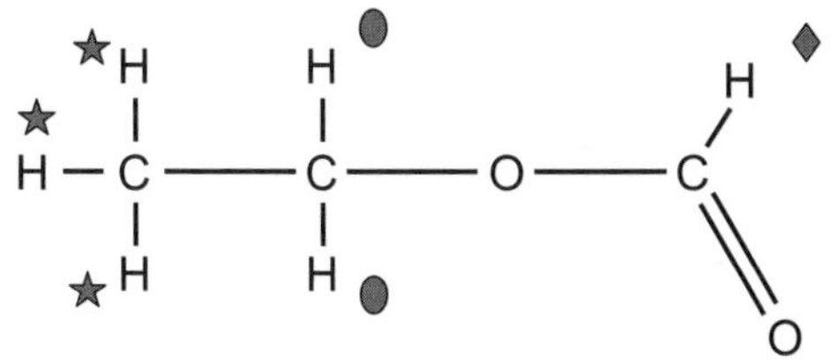

This molecule is likely to be ethyl methanoate.

The spectrum provided shows three different hydrogen environments. The splitting pattern of each of these environments should be, respectively, a triplet, a quartet and a singlet, and this matches the spectrum.

The shifts on the spectrum do not suggest a carboxylic acid, so propanoic acid is unlikely, and methyl ethanoate will have a different splitting pattern.

Explanatory notes

The spectrum of any other isomer will be different. Propanoic acid would also have three hydrogen environments, but one of those environments would have a much greater shift due to the carboxylic acid.

Methyl ethanoate would have only two hydrogen environments.

The chemical shifts listed in the Data Book will offer limited help for this question, as two of the functional groups in the correct molecule are not listed.

Mark allocation: 3 marks

- 1 mark for nominating ethyl methanoate.
- 1 mark for valid reasons for ethyl methanoate.
- 1 mark for valid reasons for rejecting other possibilities.

Question 9c.

Worked solution

The spectrum is of methyl ethanoate. Methyl ethanoate has a carbonyl bond (C=O) that leads to the absorption around 1720–1840 cm^{-1}. This ester will not have a broad band absorption around 3000–3500 cm^{-1} because it does not contain an –OH bond.

It is not propan-1-ol because the spectrum does not have a broad absorption above 3000 cm^{-1} and propan-1-ol would not have an absorption around 1750 cm^{-1}.

It is not propanoic acid because the spectrum does not include a broad absorption above 3000 cm^{-1} that the –OH (acid) bond would produce.

Explanatory notes

Methyl ethanoate contains a –C=O group but no –OH group (which is consistent with the spectrum provided).

Propan-1-ol contains an –OH group but no –C=O group (whereas the spectrum shows a C=O group).

Propanoic acid contains both –C=O and –OH (acid) groups (whereas the spectrum shows no –OH).

Mark allocation: 3 marks

- 1 mark for nominating methyl ethanoate.
- 1 mark for valid reasons for methyl ethanoate.
- 1 mark for valid reasons for rejecting other possibilities.

Question 9d.

Worked solution

The most polar molecule should have the shortest retention time if a polar solvent is used. Propan-1-ol is the most polar of the three molecules, therefore it will have the shortest retention time.

Explanatory notes

In HPLC the solvent is usually contrasting with the stationary phase, where one will be polar and the other not. If a polar solvent is used, polar substances will spend more time in the solvent and have short retention times.

Mark allocation: 2 marks

- 1 mark for nominating propan-1-ol.
- 1 mark for outlining a valid reason.

Question 10a.i.

Worked solution

The m/z ratio of the parent molecular ion is 88. The relative mass of the empirical formula, C_2H_4O, is 44. So the molecular formula must be $C_4H_8O_2$.

Explanatory notes

The m/z ratio of the parent molecular ion can be used to determine the relative molecular mass of the substance. In this case, it leads to the conclusion that the molecular formula is double the empirical formula. There is a small peak with a m/z ratio of 89, which must be due to the presence of an isotope, as a value of 89 would not be consistent with the empirical formula. Isotopes that might cause this peak include 13C or 2H.

Mark allocation: 2 marks

- 1 mark for the correct molecular formula.

- 1 mark for a valid reason relating to the parent molecular ion.

Question 10a.ii.

Worked solution

$C_4H_8O_2 + e^- \rightarrow C_4H_8O_2{}^+ + 2e^-$ or $C_4H_8O_2 \rightarrow C_4H_8O_2{}^+ + e^-$

Explanatory notes

Ionisation is achieved in a mass spectrometer by bombarding the molecule with electrons, knocking further electrons from the molecule. The molecule will have a single, positive charge. Ionisation reactions occur in the gas state, however states are not required for this question.

Mark allocation: 1 mark

- 1 mark for the correct equation.

Question 10a.iii.

Worked solution

Explanatory notes

The molecular formula matches that of a carboxylic acid or an ester.

Mark allocation: 3 marks

- 1 mark for each correct structure (up to 3 marks).

Note: There are more than three possible structures.

TIP

» The VCAA Chemistry exam generally features an instrumentation question of this nature that asks you to interpret spectra to distinguish between possible isotopes. Make sure you are clear about what each instrument is used for and the obvious spectral features to look for.

Question 10b.

Worked solution

The molecule is likely to be butanoic acid because the spectrum shows absorptions at 3000 cm^{-1} and 1700 cm^{-1}, suggesting a –OH (acid) and a C=O bond.

Explanatory notes

The infrared spectrum has a broad absorption around 3000 cm^{-1}, which suggests an –OH (acid) functional group, and the absorption around 1700 cm^{-1} is consistent with that of a C=O carbonyl group. Butanoic acid has both of these absorptions, whereas an ester will not have the broad absorption around 3000 cm^{-1}.

Mark allocation: 2 marks

- 1 mark for suggesting butanoic acid.
- 1 mark for using valid absorptions for justification.

Question 10c.

Worked solution

Butanoic acid has four hydrogen environments. The diagram below shows which part of the molecule causes each set of peaks.

From left to right, the:

- $-OH$ has no neighbouring hydrogen atoms → singlet
- first CH_2 has two hydrogen neighbours → triplet
- second CH_2 has five hydrogen neighbours → sextet
- $-CH_3$ has two hydrogen neighbours → triplet.

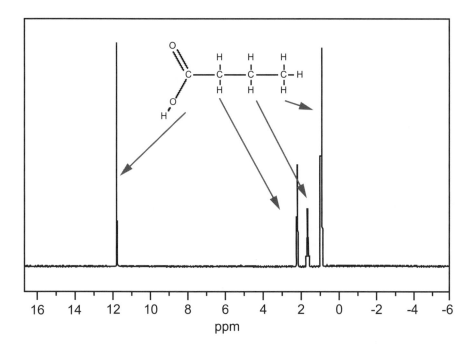

Mark allocation: 4 marks

- 1 mark for each splitting pattern identified and explained (up to 4 marks).

Note: An explanation that refers to the shift values of each environment could also achieve full marks.

TIP

> » Students often pick the correct structure in instrumentation questions but can find it difficult to explain how they know their answer is correct. The use of an annotated sketch of the molecule can be an effective way to demonstrate this understanding.

Question 11a.

Worked solution

Explanatory notes

The structures of the three amino acids can be found in the Data Book. When the amino acids join, peptide links are formed.

Mark allocation: 3 marks

- 1 mark for the correct side.
- 1 mark for the correct amide link.
- 1 mark for the correct ends to the tripeptide.

Question 11b.i.

Worked solution

Ionic bonding because –COOH is likely to donate a proton. (H-bonding can also occur.)

Mark allocation: 1 mark

- 1 mark for ionic bonding.

Question 11b.ii.

Worked solution

disulfide covalent bonds between sulfur atoms

Mark allocation: 1 mark

- 1 mark for disulfide or covalent.

Question 11b.iii.

Worked solution

dispersion forces, as hydrogen has a low value of electronegativity

Explanatory notes

Glutamic acid has a carboxyl side group, –COOH. This can donate a proton to form an ion. Cysteine contains a sulfur atom that can bond with the sulfur atom on a nearby structure. Glycine has only a hydrogen atom for a side group. This can form dispersion forces.

Mark allocation: 1 mark

- 1 mark for dispersion forces.

Question 12a.i.

Worked solution

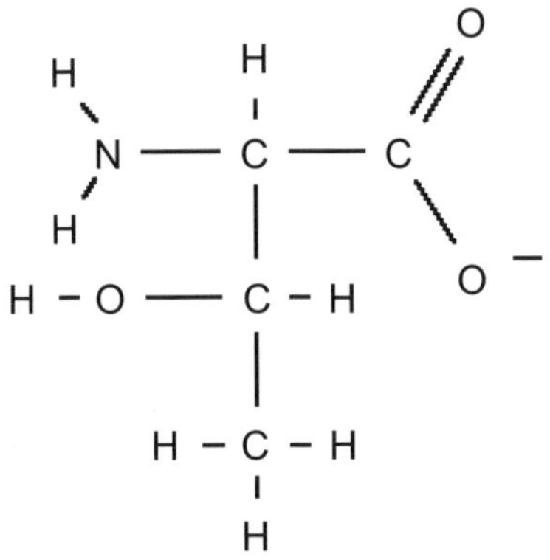

Explanatory notes

The structure of threonine is provided in the Data Book. The significance of the alkaline conditions is that the amino acid will donate a proton from its structure, leaving it as a negative ion.

Mark allocation: 1 mark

- 1 mark for the correct structure.

Question 12a.ii.

Worked solution

Threonine has two chiral carbon atoms.

Explanatory notes

The two chiral carbons are marked on the diagram below with a grey star. A chiral carbon is a carbon atom bonded to four different groups.

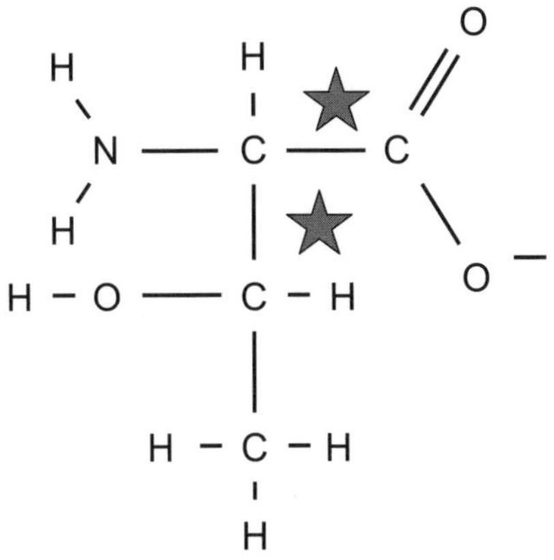

Mark allocation: 1 mark

- 1 mark for the correct answer.

Question 12b.i.

Worked solution

Ethanol and methanol are very similar, as both have a single hydroxyl group on the end of the molecule. If methanol fits an active site, it is unsurprising that ethanol might also fit.

Explanatory notes

Enzymes have very specific active site shapes. The similarity between the shape of ethanol and methanol allows both to fit the active site of this reaction.

Mark allocation: 1 mark

- 1 mark for reference to similar molecule shape.

Question 12b.ii.

Worked solution

Ethanol is a competitive enzyme inhibitor. It can occupy the enzyme site, blocking the reaction with methanol. Hence, the rate of the reaction that forms toxic chemicals is slower.

Explanatory notes

As ethanol has a similar shape and bonding to methanol, it can also occupy the dehydrogenase enzyme site. When it does, that particular enzyme will not produce any toxins and the impact of the methanol is reduced.

Mark allocation: 2 marks

- 1 mark for identifying ethanol as a competitive enzyme inhibitor.
- 1 mark for explaining that ethanol is occupying the active site.

Question 12c.i.

Worked solution

L- and D-ascorbic acid are enantiomers or optical isomers. They have the same structure but a different spatial arrangement around the chiral carbons.

Explanatory notes

Ascorbic acid has a chiral carbon atom, leading to it having two enantiomers.

Mark allocation: 1 mark

- 1 mark for reference to optical isomers.

Question 12c.ii.

Worked solution

L-ascorbic acid has the correct spatial arrangement to fit an enzyme that helps to prevent scurvy. D-ascorbic acid has a different shape and does not fit the active site, so it has no positive impact upon the body.

Mark allocation: 2 marks

- 1 mark for different shapes interacting with enzymes.
- 1 mark for the consequence of the different shapes.

Question 12d.i.

Worked solution

Gallic acid could be extracted from a plant by solvent extraction. The plant is blended, soaked in a solvent and the extract separated from the solid by filtration.

Explanatory notes

Active ingredients are usually separated by fractional distillation or solvent extraction. Solvent extraction requires blending of the plant and soaking in a solvent. The active ingredient dissolves in the solvent, allowing it to be separated from the plant.

Mark allocation: 1 mark

- 1 mark for solvent extraction.

Question 12d.ii.

Worked solution

The polarity of the solvent is important. It must match the polarity of the active ingredient that is being investigated. Gallic acid is likely to be more polar than terpineol, therefore requiring a more polar solvent.

Explanatory notes

Like dissolves like. Gallic acid contains several strong dipoles and will be a polar molecule. A polar solvent is needed to extract it. Terpineol will be relatively non-polar.

Mark allocation: 2 marks

- 1 mark for the importance of solvent polarity.
- 1 mark for comparing the polarity of the two active ingredients.

Question 12d.iii.

Worked solution

The solvent used is likely to extract many components of the plant. Chromatography is used to separate these components to prepare pure samples.

Mark allocation: 1 mark

- 1 mark for the correct answer.

Unit 4 | Area of Study 3   How is scientific inquiry used to investigate the sustainable production of energy and/or materials?

Multiple choice

Question 1

Answer: **B**

Explanatory notes

Option B is correct. The temperature is kept at 50 °C, the concentration is varied and the change in concentration causes the reaction time to change (dependent variable).

Option A is incorrect. The dependent and independent variables are the wrong way around.

Options C and D are incorrect. Both have more than one error.

Question 2

Answer: **A**

Explanatory notes

Option A is correct. Repeated trials serve to negate the impact of random errors.

Option B is incorrect. A poor choice of indicator is a mistake rather than a random error.

Option C is incorrect. It is more likely to lead to a systematic error.

Option D is incorrect. Repeating the experiment will not solve the issues caused by a systematic error.

 TIP

» **Know your error categories and memorise examples from each category. This is a common theme in questions, either in multiple-choice format or in descriptive questions involving experiment design.**

Question 3

Answer: **B**

Explanatory notes

Option B is correct. The optimum temperature for many enzymes is around 40 C, when the time for the reaction is the shortest. The graph shows this.

Options A, C and D all assume the vertical axis is reaction rate but it is not.

Short answer

Question 1a.

Worked solution

Error 1: The concentration of the three solutions is not the same. The electrochemical series assumes concentrations of 1.0 M, so all three cells should have the same concentration.

Error 2: The temperature may not have been constant if the experiments were performed over a number of days. The temperature of the laboratory should be regulated or the cells tested at the same time.

Note: Other possible answers include consideration of the purity of solutions, the nature of the salt bridges used or the need to run multiple experiments. It is not appropriate to say the results are wrong because the order of reactivity does not match that of the electrochemical series. The results are as shown – it is up to students to identify errors.

Explanatory notes

The electrochemical series is prepared under standard laboratory conditions. The temperature used is always 25 °C, the concentration is 1.0 M and the pressure of any gases is at 1 atm. Each cell needs to be constructed under these standard conditions.

Mark allocation: 4 marks

- 1 mark for each valid error (up to 2 marks).
- 1 mark for an explanation of how to fix each error (up to 2 marks).

Question 1b.

Worked solution

$Cu^{2+}(aq)$ + $2e^-$ ⇌ Cu: 0.59 V

$Co^{2+}(aq)$ + $2e^-$ ⇌ Co: −0.03 V

Explanatory notes

From the Data Book:

The difference in voltage between nickel and copper is: $0.34 - (-0.25) = 0.59$

The difference in voltage between nickel and cobalt is: $-0.28 - (-0.25) = -0.03$

Mark allocation: 2 marks

- 1 mark for each correct voltage (up to 2 marks).

Question 2a.

Worked solution

$Mg(s) + 2HCl(aq) \rightarrow MgCl_2(aq) + H_2(g)$

Explanatory notes

An acid and a metal will produce a salt and hydrogen gas.

Mark allocation: 1 mark

- 1 mark for a correct equation with states.

Question 2b.i.

Worked solution

HCl concentration

Mark allocation: 1 mark

- 1 mark for the correct answer.

Question 2b.ii.

Worked solution

time taken for the volume to reach 20 mL

Mark allocation: 1 mark

- 1 mark for the correct answer.

Question 2b.iii.

Worked solution

Possible answers include: mass of magnesium, volume of acid, temperature, volume of gas collected.

Explanatory notes

The solutions have different HCl concentrations. The time taken to produce a set amount of gas will depend upon the concentration. Many variables in this experiment are controlled, such as the amount of magnesium and the volume of acid.

Mark allocation: 2 marks

- 1 mark for each controlled variable (up to 2 marks).

Question 2c.

Worked solution

Possible answers include:

The magnesium pieces should be weighed, rather than cut to length.

The volume of acid should be measured more accurately than with a beaker.

The ability to get a stopper in the flask quickly relies upon good technique.

Explanatory notes

A number of steps in this experiment were not performed carefully. Examples: volumes should not be measured using beakers; all flasks should not be assumed to be the same; and each length of magnesium will not necessarily provide the same mass. There are a significant number of dilutions conducted. This requires care with all measurements.

Mark allocation: 2 marks

- 1 mark for each valid reason given (up to 2 marks).

Unit 4, Area of Study 3: How is scientific inquiry used to investigate the sustainable production of energy and/or materials?

185

Question 2d.

Worked solution

No. The time taken to obtain 20 mL reduces as the concentration increases, but this is not the rate of the reaction. A short time means a fast rate. The rate of reaction has increased with temperature.

Explanatory notes

As the concentration increases, the rate also increases. This is evident by the shorter time required to collect a sample. The student does not realise that the time is inversely proportional to the rate.

Mark allocation: 2 marks

- 1 mark for stating the conclusion is wrong.
- 1 mark for explaining that time is inversely proportional to rate.

Question 2e.

Worked solution

The reaction between magnesium and hydrochloric acid is highly exothermic. This means that temperature is not controlled. As the concentration increases, the effect of temperature change becomes more significant.

Explanatory notes

This is a very exothermic reaction. The temperature rise interferes with the controlled experiment environment the student is seeking. The rate increases more than expected because the temperature is so high.

Mark allocation: 2 marks

- 1 mark for nominating temperature rise.
- 1 mark for explaining that temperature will impact the reaction rate.

Question 3a.i.

Worked solution

Possible answers include: volume of water added or concentration of copper solution.

Explanatory notes

Independent variable: the extra increments of water added to the copper half-cell prior to each reading

Mark allocation: 1 mark

- 1 mark for identifying the water volume or concentration of copper solution.

Question 3a.ii.

Worked solution

the current produced by the cell

Explanatory notes

The current produced in the cell varies due to the concentration of the solution. The current depends upon the volume of water added.

Mark allocation: 1 mark

- 1 mark for identifying the cell current.

Question 3a.iii.

Worked solution

One example of a controlled variable is that the same salt bridge is retained.

Explanatory notes

There are several controlled variables: the same zinc half-cell (and zinc ion concentration) is used the whole time; the same salt bridge; the same electrodes; and the temperature can be assumed to be constant.

Mark allocation: 1 mark

- 1 mark for a valid response.

Question 3b.

Worked solution

The data suggests the opposite conclusion to that made by the student (i.e. that the current is increasing as the concentration decreases). The student has not realised that the addition of water is actually lowering the concentration of the copper half-cell.

Explanatory notes

Each water increment is reducing the copper ion concentration. The data shows that the current is increasing as the concentration drops.

Mark allocation: 2 marks

- 1 mark for stating that the student's conclusion is inconsistent with the data.
- 1 mark for outlining why the student's conclusion is inconsistent.

Unit 4, Area of Study 3: How is scientific inquiry used to investigate the sustainable production of energy and/or materials?

187

Question 3c.

Worked solution

As the student adds water, they are inadvertently introducing another variable: the surface area of the electrode in the solution. Each increment means more of the electrode is covered by solution and, perhaps, it is this increase that is leading to the increase in current. The student should control the volume of the copper solution by preparing solutions of a range of concentrations and adding the same volume of each solution each time a reading is to be taken.

Explanatory notes

The student should ensure the volume used for each copper solution is constant, so that concentration is the only variable being investigated. The preparation of a series of standard solutions of a set volume would allow this.

Mark allocation: 3 marks

- 1 mark for identifying a flaw with the experimental design.

- 2 marks for offering a changed design to address this problem.

Note: There could be other valid responses to this question, such as the need for constant stirring of the solution or a preference for using a new zinc half-cell for each experiment.